QMC
D0414614
ETURN

£4/15/-

77767
22.9.69.

APPLICATION OF DISTRIBUTIONS TO THE THEORY OF ELEMENTARY PARTICLES IN QUANTUM MECHANICS

Documents on Modern Physics

Edited by

ELLIOTT W. MONTROLL, University of Rochester
GEORGE H. VINEYARD, Brookhaven National Laboratory
MAURICE LÉVY, Université de Paris

A. H. COTTRELL, Theory of Crystal Dislocations
A. ABRAGAM, L'Effet Mössbauer
A. B. PIPPARD, The Dynamics of Conduction Electrons
K. G. BUDDEN, Lectures on Magnetoionic Theory
S. CHAPMAN, Solar Plasma, Geomagnetism and Aurora
R. H. DICKE, The Theoretical Significance of Experimental Relativity
BRYCE S. DEWITT, Dynamical Theory of Groups and Fields
J. W. CHAMBERLAIN, Motion of Charged Particles in the Earth's Magnetic Field
JOHN G. KIRKWOOD, Selected Topics in Statistical Mechanics
JOHN G. KIRKWOOD, Macromolecules
JOHN G. KIRKWOOD, Theory of Liquids
JOHN G. KIRKWOOD, Theory of Solutions
JOHN G. KIRKWOOD, Proteins
JOHN G. KIRKWOOD, Quantum Statistics and Cooperative Phenomena
JOHN G. KIRKWOOD, Shock and Detonation Waves
John G. KIRKWOOD, Dielectrics—Intermolecular Forces—Optical Rotation
HONG-YEE CHIU, Neutrino Astrophysics
M. TINKHAM, Superconductivity
DAVID HESTENES, Space-Time Algebra
P. H. E. MEIJER, Quantum Statistical Mechanics
H. REEVES, Stellar Evolution and Nucleosynthesis
J. LEQUEUX, Physics and Evolution of Galaxies
V. KOURGANOFF, Introduction to the General Theory of Particle Transfer
L. SCHWARTZ, Application of Distributions to the Theory of Elementary Particles in Quantum Mechanics
M. NIKOLIĆ, Analysis of Scattering and Decay
M. NIKOLIĆ, Kinematics and Multiparticle Systems

Additional volumes in preparation

Application of Distributions to the Theory of Elementary Particles in Quantum Mechanics

LAURENT SCHWARTZ

Université de Paris

GORDON AND BREACH

Science Publishers

NEW YORK . LONDON . PARIS

Copyright © 1968 by Gordon and Breach, Science Publishers, Inc.
150 Fifth Avenue, New York, N.Y. 10011

Library of Congress Catalog Card Number: 68-17535

Editorial Office for Great Britain:
Gordon and Breach Science Publishers Ltd.
8 Bloomsbury Way
London WC1

Editorial Office for France:
7–9 rue Emile Dubois
Paris 14

Distributed in France by:
Dunod Editeur
92 rue Bonaparte
Paris 6

Distributed in Canada by:
The Ryerson Press
299 Queen Street West
Toronto 2B, Ontario

Printed in Great Britain by Page Bros. (Norwich) Ltd.

Editors' Preface

SEVENTY years ago when the fraternity of physicists was smaller than the audience at a weekly physics colloquium in a major university a J. Willard Gibbs could, after ten years of thought, summarize his ideas on a subject in a few monumental papers or in a classic treatise. His competition did not intimidate him into a muddled correspondence with his favorite editor, nor did it occur to his colleagues that their own progress was retarded by his leisurely publication schedule.

Today the dramatic phase of a new branch of physics spans less than a decade and subsides before the definitive treatise is published. Moreover modern physics is an extremely interconnected discipline and the busy practitioner of one of its branches must be kept aware of breakthroughs in other areas. An expository literature which is clear and timely is needed to relieve him of the burden of wading through tentative and hastily written papers scattered in many journals.

To this end we have undertaken the editing of a new series, entitled *Documents on Modern Physics*, which will make available selected reviews, lecture notes, conference proceedings, and important collections of papers in branches of physics of special current interest. Complete coverage of a field will not be a primary aim. Rather, we will emphasize readability, speed of publication, and importance to students and research workers. The books will appear in low-cost paper-covered editions, as well as in cloth covers. The scope will be broad, the style informal.

From time to time, older branches of physics come alive again, and forgotten writings acquire relevance to recent developments. We expect to make a number of such works available by including them in this series along with new works.

ELLIOTT MONTROLL
GEORGE H. VINEYARD
MAURICE LÉVY

Preface

I gave lectures in 1958 in the National University of Argentine, Buenos-Aires, and in 1960, in the University of California, Berkeley, on application of distributions to the study of relativistic elementary particles. These lectures were mimeographed by the Universities concerned; both texts are now out of print. The English text has been translated into Russian (MIR Publications, Moscow 1964). This book is a revised version of the Berkeley lectures notes of 1960. Therefore it is not written and planned as a textbook. Each part was written after the lecture had been given—hence some repetitions or even modifications in the book. This text contains a large amount of pure mathematics: theory of vector valued distributions, tensor products of topological vector spaces, Fourier transforms and Bochner theorem, Hilbert sub-spaces and associated kernels have been exposed as such, but many proofs have been omitted, being intended primarily for physicists. I tried to make a distinction between scalar valued functions or distributions printed in normal characters, and vector valued, printed in bold characters. But this tedious distinction has been progressively abandoned throughout, and everything is written in normal letters, even for vectors, except in special cases where the distinction is important.

L. SCHWARTZ

Contents

CHAPTER 1

Position of the Problem

Introduction

Quantum mechanics deals with the description of motions of particles. All the information needed for the complete description of the motion of a particle is contained in its wave function $\psi(x, y, z, t)$, a complex function of position $(x, y, z) \in R^3$ (three-dimensional Euclidean space) and the time t. In non-relativistic quantum mechanics $|\psi(x, y, z, t)|^2$ is the probability density of the position of the particle. The probability of the position of a particle being in a region $A \subset R^3$ at any time t is $\iiint_A |\psi(x, y, z, t)|^2 \, dx\, dy\, dz$. Note ψ must be square integrable for each t and we assume $\iiint_{R^3} |\psi(x, y, z, t)|^2 \, dx\, dy\, dz = 1$ for each t. If we define an inner product $\iiint_{R^3} \psi_1(x, y, z, t)\overline{\psi_2(x, y, z, t)} \, dx\, dy\, dz$, then the function ψ belongs to a Hilbert space for each t.

In non-relativistic quantum mechanics, ψ satisfies the Schroedinger wave equation:

$$i\hbar\,(\partial\psi/\partial t) = H\psi$$

where $\hbar$ is Plank's constant divided by 2π, and H is a self-adjoint operator in the Hilbert space L^2 of square-integrable functions on R^3. It follows from the Schroedinger equation that the inner product of two wave functions remains constant for all time. When the particle is free of interaction,

$$H = -\frac{\hbar^2}{2m}\Delta$$

where m is the mass of the particle and Δ is the Laplacian.

In relativistic quantum mechanics, space and time are not separate; thus one cannot say that ψ is a function of four variables, unless a Lorentz coordinate system is chosen. In order to treat space and time together, the space E_4, a four-dimensional affine space, is introduced and ψ is defined on E_4. An affine space will be defined later.

Definition. A *particle*, $\mathscr{H}$, is a Hilbert space of functions on E_4.

Definition. A motion, ψ, is an element of $\mathscr{H}$ with $\|\psi\| = 1$.

Let σ be an arbitrary Lorentz transformation in E_4 and G be the Lorentz group. Under the transformation σ, a function ψ goes into a function $\sigma\psi$.

Definition. If, for all $\sigma \in G$,

$$\psi \in \mathscr{H} \Rightarrow \sigma\psi \in \mathscr{H}$$

$$\|\sigma \psi\|_{\mathscr{H}} = \|\psi\|_{\mathscr{H}}$$

then the particle, $\mathscr{H}$, is a *universal particle.* In short, a universal particle is a particle that does not change under a Lorentz transformation.

Definition. A universal particle, $\mathscr{H}$, is *elementary* if $\mathscr{H}$ contains no subspace which transforms into itself under all $\sigma \in G$, i.e., $\mathscr{H}$ is minimal.

We shall show later that the space $\mathscr{H}$ depends on a parameter $m_0 \geqslant 0$ and a parameter taking on the two values $+$ and $-$. The $\pm$ parameter is interpreted as the charge and m_0 as the rest mass of the particle.

Definition. A *meson* is a scalar elementary particle (i.e., the wave function ψ is a scalar).

For a system of two particles, the Hilbert space has the same axioms as before, except that its elements are functions on $E_4 \times E_4$. Only systems of one free particle will be dealt with in these lectures.

For the sake of generality, we shall assume that our Hilbert space is not a space of functions but a space of distributions.

We, therefore, begin with a short introduction to the theory of distributions.

Elements of the Theory of Distributions

Let R^n denote the n-dimensional Euclidean space and let $\mathscr{D}(R^n)$ (or simply $\mathscr{D}$) be the space of all complex valued functions φ defined in R^n which have derivatives of all orders and which vanish identically outside a bounded region in R^n. The functions φ will be called *testing functions.* Note that $\mathscr{D}(R^n)$ is a linear space.

We introduce now a topology in $\mathscr{D}$.

Definition. A sequence of testing functions $\{\varphi_j(x)\}$ *converges to zero in* $\mathscr{D}$ if all the functions $\varphi_j(x)$ vanish identically outside the same bounded region in R^n and if the functions $\varphi_j(x)$ and all their derivatives converge uniformly to zero.

Definition. A *distribution* T is a continuous linear functional on $\mathscr{D}$, i.e., the image under T of an element $\varphi \in \mathscr{D}$ is a complex number denoted by $\langle T, \phi \rangle$ such that

$$\langle T, (c_1\varphi_1 + c_2\varphi_2)\rangle = c_1\langle T, \varphi_1\rangle + c_2\langle T, \varphi_2\rangle$$

and

$$\varphi_j \underset{\mathscr{D}}{\rightarrow} 0 \text{ implies that } \langle T, \varphi_j\rangle \rightarrow 0.$$

Let $\mathscr{D}'(R^n)$ (or simply $\mathscr{D}'$) denote the space of distributions on R^n.

Example. Let f be a locally integrable function in R^n.
Then

$$\langle f, \varphi\rangle = \int_{R^n} f(x)\varphi(x)\,\mathrm{d}x = \int_A f(x)\varphi(x)\,\mathrm{d}x$$

defines a distribution. Here A is a bounded region in R^n (the support of φ). Thus every locally integrable function defines a distribution. Clearly, f_1 and f_2 define the same distribution if and only if $f_1 = f_2$ almost everywhere. Considering the Lebesgue classes defined by this relation (i.e., identifying functions which are equal almost everywhere), we conclude that the Lebesgue classes of locally integrable functions form a subspace of the space of distributions.

Other important examples are the Dirac distribution, δ, defined by

$$\langle \delta, \varphi\rangle = \varphi(0)$$

or

$$\langle \delta(a), \varphi\rangle = \varphi(a)$$

and the dipole ζ defined by

$$\langle \zeta, \varphi\rangle = -\varphi'(0).$$

Definition. The *derivative of a distribution* T is defined by the formula:

$$\langle T', \varphi\rangle = -\langle T, \varphi'\rangle$$

From this it follows that

$$\langle T^{(m)}, \varphi\rangle = (-1)^m \langle T, \varphi^{(m)}\rangle$$

$$\left\langle \frac{\partial T}{\partial x_k}, \varphi\right\rangle = -\left\langle T, \frac{\partial \varphi}{\partial x_k}\right\rangle$$

$$\langle D^p T, \varphi\rangle = (-1)^{|p|} \langle T, D^p \varphi\rangle$$

where p denotes the n-tuple of integers $p = (p_1, \ldots, p_n)$,

$$|p| = p_1 + \ldots + p_n$$

and

$$D^p = \left(\frac{\partial}{\partial x_1}\right)^{p_1} \left(\frac{\partial}{\partial x_2}\right)^{p_2} \cdots \left(\frac{\partial}{\partial x_n}\right)^{p_n}.$$

Thus every distribution has derivatives of all orders.

Example. Consider the Heaviside function $Y(x)$ defined by

$$Y(x) = \begin{cases} 1 & x > 0 \\ 0 & x < 0 \end{cases}$$

Then

$$\langle Y', \varphi \rangle = -\langle Y, \varphi' \rangle = -\int_{-\infty}^{\infty} Y(x)\, \varphi'(x)\, dx$$

$$= -\int_{0}^{\infty} \varphi'(x)\, dx = \varphi(0) = \langle \delta, \varphi \rangle$$

Therefore $Y' = \delta$.

Definition. Let f be a continuous function and let $A = \{x : f(x) \neq 0\}$. The closure $\overline{A}$ of A is called the *support* of the function f.

Definition. Let Ω be an open set in R^n and let $T \in \mathscr{D}'$. We say that $T = 0$ in Ω if $\langle T, \varphi \rangle = 0$ for all $\varphi \in \mathscr{D}$ whose support is contained in Ω. For example, $\delta = 0$ in $R - \{0\}$.

Theorem. Let $\{\Omega_i\}$ be any system of open subsets in R^n and suppose that $T = 0$ in every Ω_i. Then $T = 0$ in $\cup \Omega_i$.

Proof. We must show that $\langle T, \varphi \rangle = 0$ for every $\varphi \in \mathscr{D}$ whose support is contained in $\cup \Omega_i$. Let A be the support of some $\varphi \in \mathscr{D}$. Since A is compact and covered by $\{\Omega_i\}$, there exists a finite subcover $\{\Omega_{i_k}\}$, $k = 1, \ldots, n$. Let $\{\psi_k\}$, $k = 1, \ldots, n$, be an infinitely many times continuously differentiable partition of unity on A with respect to Ω_{i_k}, that is, $\psi_k \in \mathscr{D}(R^n)$, each ψ_k has its support in Ω_{i_k} and

$$\sum_{k=1}^{n} \psi_k = 1$$

on A. Then

$$\langle T, \varphi \rangle = \langle T, \sum_{k=1}^{n} \psi_k \varphi \rangle = \sum_{k=1}^{n} \langle T, \psi_k \varphi \rangle = 0.$$

Corollary. For every distribution T there exists exactly one maximal open subset of R^n in which T is zero.

Proof. Consider all Ω_i in which $T = 0$. Then $\cup \Omega_i$ is the required set.

Definition. The *support* of T is the complement of the maximal open subset of R^n in which $T = 0$.

We introduce now a topology in the space of distribution $\mathscr{D}'$. Since it is a linear space it suffices to define convergence to zero.

Definition. *Weak convergence*: Let $\{T_j\}$ be a sequence in $\mathscr{D}'$. We say that T_j converges to zero in the sense of distributions, or $T_j \to 0$ in $\mathscr{D}'$, if $\langle T_j, \varphi \rangle \to 0$ for every $\varphi \in \mathscr{D}$.

Strong convergence requires a certain uniformity and it will be defined when needed.

Theorem. Differentiation is a continuous operation, i.e. $T_j \to 0$ in $\mathscr{D}'$ implies that $T_j' \to 0$ in $\mathscr{D}'$.

Proof. $\langle T_j', \varphi \rangle = -\langle T_j, \varphi' \rangle \to 0$ for every $\varphi \in \mathscr{D}$.

Remarks. The weak topology defined here makes convergent a lot of sequences which are ordinarily divergent. A series which is convergent in the sense of distributions may be differentiated term by term, i.e., if $T = \sum T_j$ then $T' = \sum T_j'$.

Theorem. Let $f_j \to 0$ almost everywhere and suppose that $|f_j| \leqslant g$, where g is a fixed positive locally integrable function. Then $f_j \to 0$ in the sense of distributions.

Proof. This follows from the Lebesgue convergence theorem.

Example. A trigonometric series

$$\sum_k a_k \, e^{2\pi i k x}$$

is convergent in the sense of distributions if and only if $|a_k| \leqslant A k^\alpha$, for $k \neq 0$, where A is a constant and α is some positive integer. Thus many trigonometric series become convergent in the sense of distributions. To see this consider the series

$$\sum_{k \neq 0} \frac{a_k}{(2\pi i k)^{\alpha+2}} \, e^{2\pi i k x}.$$

It is uniformly convergent since

$$\frac{|a_k|}{|2\pi i k|^{\alpha+2}} \leqslant \frac{A}{(2\pi)^{\alpha+2}} \frac{1}{k^2}.$$

Therefore this series converges also in the sense of distributions. If we differentiate now $\alpha + 2$ times term by term we obtain the original series which therefore is convergent in the sense of distributions.

Examples. The series

$$\sum_{k=-\infty}^{\infty} e^{2\pi ikx}$$

is ordinarily divergent. However, in the sense of distributions it converges to the distribution

$$\sum_{k=-\infty}^{\infty} \delta(x-k)$$

δ	δ	δ	δ	δ
-2	-1	0	1	2

Differentiating term by term we see that

$$\sum_{-\infty}^{\infty} (2\pi ik)\, e^{2\pi ikx}$$

converges to

$$\sum_{k=-\infty}^{\infty} \delta'(x-k)$$

δ'	δ'	δ'	δ'	δ'
-2	-1	0	1	2

Affine Spaces: Lorentz Transformations

In the previous section we defined the space $\mathscr{D}'(R^n)$ of distributions on the Euclidean space R^n. In a similar way we may define the space $\mathscr{D}'(\mathbf{E}_n)$ of distributions on the n-dimensional vector space $\mathbf{E}_n$. However, in physical space there is no pre-determined origin, so that we do not have an $\mathbf{E}_n$ to start with. For this reason, we introduce the concept of an affine space.

Definition. An *affine space* is a set E and an associated vector space $\mathbf{E}$. This association is defined by a map from $E \times E$ to $\mathbf{E}$ which maps a pair a, b of elements of E to the vector $\mathbf{ab}$ of E, and such that the following two laws are satisfied:

(1) *Chasles' relation*: If a, b, c are any three elements of E, then $\mathbf{ab} + \mathbf{bc} + \mathbf{ca} = 0$.

(2) Let o be a fixed element of E. The map $a \to \mathbf{oa}$ is a one-to-one correspondence between E and $\mathbf{E}$.

It should be noted that (1) may be generalized to more than three elements. Furthermore, according to (1) the triple a,a,a yields 3 $\mathbf{aa} = 0$ or $\mathbf{aa} = 0$ and the triple a,a,b yields $\mathbf{ab} + \mathbf{ba} = 0$.

For obvious reasons, the notation

$$\mathbf{ab} = \overrightarrow{b - a}$$

is very convenient. Thus the difference between two elements a,b of E is a map which maps the pair a,b to the vector $\mathbf{ab}$ of $\mathbf{E}$, and which obviously satisfies the above two laws. If a is a given element of E and $\mathbf{x}$ is a given element of $\mathbf{E}$, then there exists one and only one element $b \in E$ such that $a + \mathbf{x} = b$ where this equality is equivalent to $\mathbf{x} = \overrightarrow{b - a}$.

Definition. Let E and F be two affine spaces. The map

$$\sigma : E \rightarrow F$$

is called an *affine operator* from E to F if there exists an associated linear operator

$$\boldsymbol{\sigma} : \mathbf{E} \rightarrow \mathbf{F}$$

such that

$$\overrightarrow{\sigma b - \sigma a} = \boldsymbol{\sigma}(\overrightarrow{b - a}).$$

Note that the associated linear operator $\boldsymbol{\sigma}$ is uniquely determined by σ. Furthermore, the composition of two affine operators is an affine operator and the invertible affine operators form a group.

Example. The translation $\mathbf{U} : x \rightarrow x + \mathbf{U}$ is an affine operator from the affine space E onto itself. The associated linear operator of a translation is the identity operator,

$$\overrightarrow{(b + \mathbf{U}) - (a + \mathbf{U})} = \overrightarrow{b - a}$$

Conversely, every affine operator having the identity as its associated linear operator is a translation.

Let $\mathbf{E}$ be a vector space over the reals and consider a quadratic form $(\mathbf{x}|\mathbf{y})$ defined on $\mathbf{E}$. It is assumed that $(\mathbf{x}|\mathbf{y})$ is bilinear, symmetric $((\mathbf{x}|\mathbf{y}) = (\mathbf{y}|\mathbf{x}))$ and non-degenerate (no element except zero is orthogonal to the whole space).

Let $\mathbf{e}_1, \mathbf{e}_2, \ldots, \mathbf{e}_n$ be an orthonormal basis in $\mathbf{E}$, i.e., $(\mathbf{e}_i|\mathbf{e}_j) = 0$ for $i \neq j$ and $(e_i|e_i) = \pm 1$. Every finite dimensional vector space with a non-degenerate quadratic form has an infinite number of orthonormal bases. However, the number of basis elements e such that $(e|e) = +1$

and the number of basis elements e such that $(e|e) = -1$ is independent of the particular chosen basis.

Definition. The *signature* of an n-dimensional vector space with respect to a given quadratic form $(x|y)$ is the pair of integers (p, q), where $p + q = n$, p is the number of O.N. basis elements e such that $(e|e) = +1$ and q is the number of O.N. basis elements e such that $(e|e) = -1$.

Definition. A *Lorentz four-dimensional vector space* is a vector space with a quadratic form which has the signature (3,1). The orthonormal basis will be denoted by $\mathbf{e}_1$, $\mathbf{e}_2$, $\mathbf{e}_3$, $\mathbf{e}_0$, where $(e_i|e_i) = +1$, $i = 1,2,3$ and $(\mathbf{e}_0|\mathbf{e}_0) = -1$. A *Lorentz four-dimensional affine space* is an affine space E_4 whose associated vector space $\mathbf{E}_4$ has the signature (3,1). By a *Galilean reference system* we mean a chosen origin 0 in E_4 and a chosen orthonormal basis $\mathbf{e}_1$, $\mathbf{e}_2$, $\mathbf{e}_3$, $\mathbf{e}_0$ in $\mathbf{E}_4$ (chosen coordinate system).

Every point of the universe has four coordinates x_1, x_2, x_3, $x_0 = ct$, three of space and one of time.

Definition. A *Lorentz transformation* σ is an affine invertible operator in a Lorentz affine space which preserves its Lorentz structure, i.e., the associated linear operator preserves the quadratic form

$$(\boldsymbol{\sigma}\mathbf{x}|\boldsymbol{\sigma}\mathbf{y}) = (\mathbf{x}|\mathbf{y}).$$

The Lorentz transformations form a group. The group G consisting of all the Lorentz transformations σ will be called the *inhomogeneous Lorentz group*, whereas the group $\mathbf{G}$ consisting of the associated linear operators $\boldsymbol{\sigma}$ will be called the *homogeneous Lorentz group*.

Example. Translations are Lorentz transformations.

One may now define the space $\mathscr{D}'(E)$ of distributions over the affine space E. More generally, one can define the space $\mathscr{D}'$ for a manifold V, as the space of infinitely differentiable functions with compact support, with a suitable topology, and $\mathscr{D}'(V)$, space of distributions on V, as its dual.

Universal Scalar Particles

Now the definitions of a scalar particle and a universal particle will be made more precise.

Definition. A *scalar particle* in the universe E_4 is a set $\mathscr{H}$ satisfying the postulates:

(1) $\mathscr{H}$ is a vector subspace of $\mathscr{D}'(E_4)$.

(2) $\mathscr{H}$ is equipped with a Hilbertian structure, that is, there is a linear-antilinear form $(\psi_1|\psi_2)_{\mathscr{H}}$ (linear in ψ_1 and antilinear in ψ_2) in $\mathscr{H}$ which is Hermitian and positive definite, and $\mathscr{H}$ is complete with respect to the norm $\|\psi\|_{\mathscr{H}} = (\psi|\psi)^{\frac{1}{2}}_{\mathscr{H}}$.

(3) The canonical embedding of $\mathscr{H}$ into $\mathscr{D}'$ is continuous, that is,

$$\psi_j \to 0 \text{ in } \mathscr{H} \Rightarrow \psi_j \to 0 \text{ in } \mathscr{D}'.$$

We shall find that $\mathscr{H}$ represents charged particles. If the distributions in E_4 were restricted to be real valued, then $\mathscr{H}$ would describe a neutral particle.

Definition. A motion of a particle is an element $\psi \in \mathscr{H}$ such that $\|\psi\|_{\mathscr{H}} = 1$.

A universal particle (universal with respect to the Lorentz group) is one which is considered the same by different observers. An observer makes his observations in some frame of reference; thus the particle $\mathscr{H}$ is interpreted by him as being a space of distributions over R^4 instead of E_4. If all observers interpret $\mathscr{H}$ to be the same space of distributions over R^4, then $\mathscr{H}$ is a universal particle. A more precise definition is given after the operation of $\sigma \in G$ on distributions is defined.

A Lorentz transformation $\sigma \in G$ not only operates on E_4 but also on every structure given over E_4. If $\varphi(x)$ for $x \in E_4$ is a complex function on E_4, the transformation $\varphi \to \sigma\varphi$ is defined by the equation

$$\sigma\varphi(\sigma x) = \varphi(x) \qquad x \in E_4$$

or, equivalently,

$$\sigma\varphi(y) = \varphi(\sigma^{-1}y) \qquad y \in E_4.$$

From the fact that $\boldsymbol{\sigma} \in \mathbf{G}$ is a linear operator, it follows that:

Theorem. $\varphi \in \mathscr{D}(E_4) \Rightarrow \sigma\varphi \in \mathscr{D}(E_4)$. It follows from the definition of $\sigma\varphi$ that:

Theorem. $\varphi_n \to 0 \Rightarrow \sigma\varphi_n \to 0$.

Thus σ gives an automorphism of $\mathscr{D}$ onto $\mathscr{D}$.

The operation of σ on distributions is defined by the equation

$$\langle \sigma T, \sigma\psi \rangle = \langle T, \varphi \rangle$$

or, equivalently,

$$\langle \sigma T, \psi \rangle = \langle T, \sigma^{-1}\psi \rangle = \langle T_y, \psi(\sigma y).$$

Theorem. The operator σ operates linearly and continuously on distributions.

Proof. Linearity:

$$\begin{aligned}\langle\sigma(a_1T_1 + a_2T_2), \psi\rangle &= \langle a_1T_1 + a_2T_2, \sigma^{-1}\psi\rangle\\ &= a_1\langle T_1, \sigma^{-1}\psi\rangle + a_2\langle T_2, \sigma^{-1}\psi\rangle\\ &= a_1\langle \sigma T_1, \psi\rangle + a_2\langle \sigma T_2, \psi\rangle = \langle a_1\sigma T_1 + a_2\sigma T_2, \psi\rangle.\end{aligned}$$

Continuity: Given $T_n \to 0$, then for any $\psi \in \mathscr{D}$, we have $\sigma^{-1}\psi \in \mathscr{D}$ and

$$\langle\sigma T_n, \psi\rangle = \langle T_n, \sigma^{-1}\psi\rangle \to 0$$

therefore $\sigma T_n \to 0$.

It is simple to show that the operation of σ followed by τ on $\mathscr{D}'$ is the same as the operation of $\tau\sigma$ on $\mathscr{D}'$. Then it follows that σ is an automorphism of $\mathscr{D}'$ onto $\mathscr{D}'$.

Given an affine space E and a positive measure on $\mathbf{E}$ which is invariant under translation, a measure on E is uniquely defined.

Then any locally integrable function f on E defines a distribution

$$\langle f, \varphi\rangle = \int f(x)\,\varphi(x)\,\mathrm{d}x.$$

Given a quadratic form on $\mathbf{E}$, there corresponds orthonormal bases and a Haar measure. In view of the fact that any $\boldsymbol{\sigma} \in \mathbf{G}$ preserves the quadratic form, it will also preserve the Haar measures. It follows that σ preserves the correspondence between functions and distributions for this measure.

Given a scalar particle $\mathscr{H} \subset \mathscr{D}'(E_4)$ and any $\sigma \in G$, one may form the space $\sigma\mathscr{H}$, the set of $\sigma\psi$ for all $\psi \in \mathscr{H}$. With the inner product

$$(\sigma\psi_1|\sigma\psi_2)_{\sigma\mathscr{H}} = (\psi_1|\psi_2)_{\mathscr{H}}$$

the space $\sigma\mathscr{H}$ is also a Hilbert space.

Definition. A scalar particle $\mathscr{H}$ is *universal* if for all $\sigma \in G$ the following is true:
(1) $\sigma\mathscr{H} = \mathscr{H}$
(2) $\|\sigma\psi\|_{\mathscr{H}} = \|\psi\|_{\mathscr{H}}$ for all $\psi \in \mathscr{H}$.
It follows that $\mathscr{H}$ is a universal particle if and only if every $\sigma \in G$ is a unitary operator of $\mathscr{H}$ onto $\mathscr{H}$.

Scalar and Vector Particles in an Arbitrary Universe

Definition. A *universe* V is a C^∞-manifold of finite dimension n. A group G whose elements operate on V will be called the structure group of the universe.

Definition. A *scalar particle in the universe* V is a set $\mathscr{H}$ satisfying the postulates:
(1) $\mathscr{H}$ is a vector subspace of $\mathscr{D}'(V)$, the space of distributions in V.
(2) $\mathscr{H}$ is equipped with a Hilbertian structure.
(3) $\psi_j \to 0$ in $\mathscr{H} \Rightarrow \psi_j \to 0$ in $\mathscr{D}'(V)$.

Definition. A scalar particle $\mathscr{H}$ in the universe V is *universal* (with respect to G) if for all $\sigma \in G$:
(1) $\sigma \mathscr{H} = \mathscr{H}$
(2) $\|\sigma\psi\|_{\mathscr{H}} = \|\psi\|_{\mathscr{H}}$ for all $\psi \in \mathscr{H}$.

Example. For one scalar particle, we may take $V = E_4$ with a given Lorentz quadratic form and the corresponding Lorentz group as the structure group.

Example. For two particles, we take $V = E_4 \times E_4$. The structure group G is again the Lorentz group acting on $E_4 \times E_4$ as follows: For $(x, y) \in E_4 \times E_4$ and $\sigma \in G$

$$(x, y) \to \sigma(x, y) = (\sigma x, \sigma y).$$

In order to treat particles such as the electron, proton, etc., we must introduce the concept of a vector-valued distribution. Let $\mathbf{F}$ be a finite dimensional vector space over C.

Definition. *An* $\mathbf{F}$*-valued distribution* $\mathbf{T}$ *on* V is a continuous linear map $\mathbf{T}: \varphi \to \langle \mathbf{T}, \varphi \rangle$ of $\mathscr{D}(V)$ into $\mathbf{F}$.

The space $\mathscr{D}'(V; \mathbf{F})$ of $\mathbf{F}$-valued distributions on V, the space $\mathscr{L}(\mathscr{D}(V); \mathbf{F})$ of continuous linear maps of $\mathscr{D}(V)$ into $\mathbf{F}$, and the tensor product $\mathscr{D}'(V) \otimes \mathbf{F}$ of $\mathscr{D}'(V)$ and $\mathbf{F}$ are all identical:

$$\mathscr{D}'(V; \mathbf{F}) = \mathscr{L}(\mathscr{D}(V); \mathbf{F}) = \mathscr{D}'(V) \otimes \mathbf{F}.$$

Example. Let $V = R^n$ be an affine space with a Lebesgue measure. If $\mathbf{f}(x)$ is a locally integrable $\mathbf{F}$-valued function on R^n, then to $\mathbf{f}$ corresponds a distribution

$$\varphi \to \langle \mathbf{f}, \varphi \rangle = \int \mathbf{f}(x)\, \varphi(x)\, dx.$$

If $S \in \mathscr{D}'(V)$ and $\mathbf{f} \in \mathbf{F}$, then the vector-valued distribution $S\mathbf{f} \in \mathscr{D}'(V; \mathbf{F})$ may be defined by the equation

$$\langle S\mathbf{f}, \varphi \rangle = \langle S, \varphi \rangle \mathbf{f}.$$

$S\mathbf{f}$ is identified with $S \otimes \mathbf{f} \in \mathscr{D}'(V) \otimes F$.

If $\mathbf{F}$ has the basis

$$\mathbf{f}_1, \mathbf{f}_2, \ldots, \mathbf{f}_n$$

then $\mathbf{T} \in \mathscr{D}'(V; \mathbf{F})$ can be written

$$\mathbf{T} = T_1\mathbf{f}_1 + T_2\mathbf{f}_2 + \ldots + T_n\mathbf{f}_n$$

where $T_1, T_2, \ldots, T_n \in \mathscr{D}'(V)$. Thus for $\varphi \in \mathscr{D}(V)$:

$$\langle \mathbf{T}, \varphi \rangle = \sum_{i=1}^{n} \langle T_1, \varphi \rangle \mathbf{f}_i.$$

Definition. An $\mathbf{F}$-*valued particle in the universe* V is a set $\mathscr{H} \subset \mathscr{D}'(V; \mathbf{F})$ satisfying the same postulates as a scalar particle in the universe V except that $\mathscr{D}'(V)$ is replaced by $\mathscr{D}'(V; \mathbf{F})$ in the definition and the following additional postulates are satisfied:

(1) Every $\sigma \in G$ operates not only on V, but also on $\mathbf{F}$, thus $x \in V \Rightarrow \sigma x \in V$, and $\mathbf{f} \in \mathbf{F} \Rightarrow \sigma \mathbf{f} \in \mathbf{F}$.

(2) If σ defines the identity operation in both V and $\mathbf{F}$, then σ is the identity of G.

Remark. G operates faithfully on the product $V \otimes \mathbf{F}$, but not necessarily on V or $\mathbf{F}$ alone.

Example. For an electron, G is the proper spinor group, $V = E_4$, and $\mathbf{F}$ is a two-dimensional vector space over C. There is a mapping $\sigma \in G \to \sigma_0 \in$ proper inhomogeneous Lorentz group such that two elements of G correspond to each element in the Lorentz group, and the action of each σ on any element of E_4 is the same as the action of the corresponding σ_0 given by the mapping. There is also a mapping $\sigma \in G \to \tau \in$ the set of unimodular operators in $\mathbf{F}$, which form a group, such that an infinite number of elements of G correspond to each element of this group of linear operators in $\mathbf{F}$, and the action of each σ on any element of $\mathbf{F}$ is the same as the action of the corresponding τ given by the mapping.

Definition. The operation of σ on $\mathbf{T} \in \mathscr{D}'(V) \otimes F$ is defined by the equation

$$\sigma(\langle \mathbf{T}, \varphi \rangle) = \langle \sigma \mathbf{T}, \sigma \varphi \rangle,$$

or, equivalently,

$$\langle \sigma \mathbf{T}, \psi \rangle = \sigma(\langle \mathbf{T}, \sigma^{-1} \psi \rangle) = \tau(\langle \mathbf{T}, \psi(\sigma_0 x) \rangle).$$

Definition. A *universal* $\mathbf{F}$-valued particle in the universe V is defined in exactly the same way as a *universal* scalar particle in the universe V.

Weak and Strong Convergence

Definition. Let E be a topological vector space. A set $A \subset E$ is called *convex* if whenever $x, y \in A$ the elements $ax + (1 - a)y, 0 \leqslant a \leqslant 1$, also belong to A. E is called *locally convex* if its topology can be defined by a base consisting of convex sets.

Let E be a locally convex topological vector space and let E' be its dual, i.e., the space of continuous linear forms on E. We shall define weak and strong convergence in E'.

Definition. The sequence $\{e'_j\} \subset E'$ *converges weakly* to zero, $e'_j \to 0$ weakly, if $\langle e'_j, e\rangle \to 0$ for every $e \in E$. Here the inner product is the one defined naturally as being equal to the value of e'_j at e.

Strong convergence requires a certain uniformity on the bounded subsets of E.

Definition. A subset A of E is called *bounded* if it can be mapped into any neighborhood of zero by a contraction with a non-zero ratio. For example, if E is a Banach space, a subset of E is bounded if it can be mapped into any ball by a contraction with a non-zero ratio.

Definition. A sequence $\{e'_j\} \subset E'$ *converges strongly* to zero, written $e'_j \to 0$ strongly, if $\langle e'_j, e\rangle \to 0$ for every $e \in E$ and this convergence is uniform on every bounded subset of E.

Let us return now to the spaces $\mathscr{D}(V)$ and $\mathscr{D}'(V)$. $\mathscr{D}(V)$ is the space of testing functions defined on the universe V. If K is a compact subset of V, let $\mathscr{D}_K(V)$ denote the space of testing functions whose support is in K. In $\mathscr{D}_K(V)$ we may introduce the norms

$$\|\varphi\|_m = \sup_{\substack{x \in K \\ |p| \leqslant m}} |D^p\varphi(x)|$$

where D^p denotes the $(p_1, p_2, \ldots, p_k)$ derivative. Just as before, we define convergence to zero of a sequence $\{\varphi_n\}$ in $\mathscr{D}_K(V)$ by requiring that $\|\varphi_n\|_m \to 0$ for all m.

An element T of $\mathscr{D}'(V)$ is a linear form on $\mathscr{D}(V)$ which is continuous on every $\mathscr{D}_K(V)$. A sequence $\{T_j\} \subset \mathscr{D}'(V)$ *converges weakly* to zero, $T_j \to 0$ weakly, if $\langle T_j, \varphi\rangle \to 0$ for every $\varphi \in \mathscr{D}(V)$. It *converges strongly* to zero, $T_j \to 0$ strongly, if $\langle T_j, \varphi\rangle \to 0$ for every $\varphi \in \mathscr{D}(V)$, and this convergence is uniform on the bounded subsets of $\mathscr{D}_K(V)$ for any K.

We state here without proof the following important theorem:

Theorem. The space $\mathscr{D}'(V; \mathbf{F})$ of $\mathbf{F}$-valued distributions on V is a locally convex topological vector space which is complete under the strong topology.

From this point on, our basic purpose is to find all the subspaces $\mathscr{H}$ of $\mathscr{D}'(V; \mathbf{F})$, such that $\mathscr{H}$ may be equipped with a Hilbertian structure and convergence in $\mathscr{H}$ implies convergence in $\mathscr{D}'(V; \mathbf{F})$.

CHAPTER 2

The Set $\mathfrak{H}$ of the Universal Particles and its Structure

The Space $\mathfrak{H}$

Let E be a complete locally convex topological vector space. In our particular study, E will be $\mathscr{D}'(V; \mathbf{F})$. Let $\mathfrak{H}$ denote the set of pairs $\{\mathscr{H}, (|)_{\mathscr{H}}\}$ consisting of a linear subspace $\mathscr{H}$ of E and a scalar product on $\mathscr{H}$ satisfying the following conditions:

(a) Provided with $(|)_{\mathscr{H}}$, $\mathscr{H}$ is an Hilbert space;

(b) The injection of $\mathscr{H}$ into E is continuous; that is, convergence in $\mathscr{H}$ implies convergence in E.

On $\mathfrak{H}$ we may define the following:

(1) *Multiplication by non-negative scalars*, $\lambda \geqslant 0$; the set

$$\lambda\mathscr{H} = \begin{cases} \mathscr{H} \text{ if } \lambda > 0 \\ 0 \text{ if } \lambda = 0 \end{cases}$$

If $T \in \mathscr{H}$ and therefore $T \in \lambda\mathscr{H}$, then

$$\|T\|_{\lambda\mathscr{H}} = 1/\sqrt{\lambda}\,\|T\|_{\mathscr{H}}$$

(2) *Addition*: The set

$$\mathscr{H}_1 + \mathscr{H}_2 = \{T : T = T_1 + T_2, \quad T_1 \in \mathscr{H}_1, \quad T_2 \in \mathscr{H}_2\}$$

with the norm

$$\|T\|_{\mathscr{H}_1 + \mathscr{H}_2} = \inf \sqrt{(\|T_1\|^2{}_{\mathscr{H}_1} + \|T_2\|^2{}_{\mathscr{H}_2})}.$$

(3) *Order*: A partial ordering is defined in $\mathfrak{H}$ by the relation $\mathscr{H}_1 \leqslant \mathscr{H}_2$ if $\mathscr{H}_1 \subset \mathscr{H}_2$ and the norm in $\mathscr{H}_1$ is greater than or equal to the norm in $\mathscr{H}_2$.

(4) *Topology*: It will be shown that $\mathfrak{H}$ is a closed convex cone in a topological vector space which we shall construct.

Definition. An *anti-kernel* L is an anti-linear continuous map of the dual space E' into E, that is, $E \overset{L}{\leftarrow} E'$, L is continuous and $L(\lambda e') = \bar{\lambda}L(e')$. L is called *positive* if $\langle e', Le' \rangle \geqslant 0$ for all $e' \in E'$. The duality product between E and E' is defined by

$\langle e', f \rangle = e'(f)$, the value of e' at f, for $f \in E$, $e' \in E'$.

Example. Let $E = C^n$, the n-dimensional complex vector space, then $E' = C^n$. An anti-kernel here is a positive definite hermitian matrix L,

$$L : (\xi_1, \ldots, \xi_n) \to (\eta_1, \ldots, \eta_n)$$

where

$$\eta_i = \sum_j L_{ij}\bar{\xi}_j,$$

and

$$\langle \xi, L\xi \rangle = \xi_1\eta_1 + \ldots + \xi_n\eta_n = \sum_{ij} L_{ij}\xi_i\bar{\xi}_j$$

In the space of positive anti-kernels addition and scalar multiplication are defined in the obvious manner. Furthermore, an order relation is given by

$$L_1 \leqslant L_2 \text{ if } \langle e', L_1 e' \rangle \leqslant \langle e', L_2 e' \rangle \text{ for all } e' \in E'.$$

We prove now the following fundamental result:

Theorem. There is a one-to-one correspondence between the elements $\mathcal{H}$ of $\mathfrak{H}$ and the positive anti-kernels L. To $\mathcal{H} \in \mathfrak{H}$ corresponds the kernel $L = J \,.\, i_{\mathcal{H}} \,.\, {}^tJ$, where J is the natural injection $\mathcal{H} \to E$, ${}^tJ$ its transposed $E' \to \mathcal{H}'$ and $i_{\mathcal{H}}$ the canonical anti-isomorphism $\mathcal{H}' \to \mathcal{H}$.

Proof. First we show that to a given $\mathcal{H}$ corresponds a positive anti-kernel L. Let $e' \in E'$. Since e' is a continuous linear functional on E and the injection of $\mathcal{H}$ into E is continuous, it follows that e' is a continuous linear functional on $\mathcal{H}$. By the Riesz representation theorem there is a unique element of $\mathcal{H}$ which we shall denote by Le', such that

$$\langle e', h \rangle = (h|Le')_{\mathcal{H}}, \quad h \in \mathcal{H}, \quad e' \in E'. \tag{2.1}$$

Clearly the map $L : E' \to \mathcal{H} \subset E$ which is defined by equation (2.1) is anti-linear. To show that L is continuous, let $e'_j \to 0$ in E' (strong topology), i.e., $\langle e'_j, h \rangle \to 0$ for every $h \in E$ and uniformly for h on bounded subsets of E. Since the injection of $\mathcal{H}$ into E is continuous, it follows that bounded subsets of $\mathcal{H}$ are bounded in E. Hence $\langle e'_j, h \rangle = (h|Le'_j) \to 0$ uniformly on the unit ball of $\mathcal{H}$. Therefore $\|Le'_j\|_{\mathcal{H}} \to 0$

and L is continuous. Finally, if we let $h = Lf'$ in equation (2.1) we obtain

$$\langle e', Lf' \rangle = (Lf' | Le')_{\mathscr{H}}, \quad e' f' \in E', \tag{2.2}$$

and if $e' = f'$, then

$$\langle e', Le' \rangle = (Le' | Le')_{\mathscr{H}} \geqslant 0, \quad e' \in E'$$

which shows that L is positive.

Conversely, we must show that to a given positive anti-kernel L corresponds an element $\mathscr{H}$ of $\mathfrak{H}$. Let $\mathscr{H}_0. = L(E') \subset \mathscr{H}$. According to equation (2.2), we define in $\mathscr{H}_0$. the inner product

$$(u|v)_{\mathscr{H}_0} = (Le' | Lf')_{\mathscr{H}_0} = \langle f', Le' \rangle, u, v \in \mathscr{H}_0 \tag{2.3}$$

where $u = Le'$, $v = Lf'$. We prove now for $\mathscr{H}_0$ the following:

(a) Definition (2.3) of $(u|v)_{\mathscr{H}_0}$ is unique, i.e., independent of the choice of e' and f' such that $u = Le'$ and $v = Lf'$. This follows immediately from equation (2.3) and the hermitian symmetry by noting that $(u|v)_{\mathscr{H}_0} = 0$ if either u or v is zero.

(b) The form $(u|v)_{\mathscr{H}_0}$ is positive definite. Since L is positive,

$$(u|u)_{\mathscr{H}_0} = \langle e', Le' \rangle \geqslant 0, \quad u \in \mathscr{H}_0.$$

If $\langle e', Le' \rangle = 0$, then by Schwarz's inequality,

$$|\langle f', Le' \rangle| \leqslant \langle f', Lf' \rangle^{\frac{1}{2}} \langle e', Le' \rangle^{\frac{1}{2}} = 0$$

for all $f' \in E'$. Hence $Le' = u = 0$ and $(u|u)_{\mathscr{H}_0} = 0$ if and only if $u = 0$.

(c) The topology of $\mathscr{H}_0$ is finer than that of E, i.e., the injection of $\mathscr{H}_0$ into E is continuous. It suffices here to show that the unit ball of $\mathscr{H}_0$, $\{Le' : \langle e', Le' \rangle \leqslant 1\}$, is a bounded subset of E. From Schwarz's inequality

$$|\langle f', Le' \rangle| \leqslant \langle f', Lf' \rangle^{\frac{1}{2}} \langle e', Le' \rangle^{\frac{1}{2}} \leqslant \langle f', Lf' \rangle^{\frac{1}{2}}$$

it follows that the set $\{\langle f', Le' \rangle : \langle e', Le' \rangle \leqslant 1, e' \in E\}$ is bounded in $\mathfrak{C}$ for each $f' \in E'$, or that the unit ball of $\mathscr{H}_0$ is weakly bounded in E. Using now Mackey's theorem, which states that a subset of a locally convex topological vector space is strongly bounded if and only if it is weakly bounded, it follows that the unit ball of $\mathscr{H}_0$ is bounded in E.

We have shown up to now that $\mathscr{H}_0 = LE'$ is a pre-Hilbert space whose injection into E is continuous. We expect to obtain the required

$\mathscr{H}$ corresponding to L by completing $\mathscr{H}_0$. It is necessary therefore to prove the following:

(1) If there exists an $\mathscr{H}$ corresponding to L such that equation (2.1) is satisfied, then $\mathscr{H}_0$ is dense in $\mathscr{H}$. Let $h \in \mathscr{H}$ such that $(h|Le')_{\mathscr{H}} = 0$ for all $e' \in E'$. Then $(h|Le')_{\mathscr{H}} = \langle e', h\rangle = 0$ for all $e' \in E'$ and, by Hahn-Banach theorem, $h = 0$. Thus an element of $\mathscr{H}$ which is orthogonal to every element of $\mathscr{H}_0$ is zero and hence $\mathscr{H}_0$ is dense in $\mathscr{H}$.

(2) The completion $\hat{\mathscr{H}}_0$ of $\mathscr{H}_0$ can be embedded in E. Consider the (continuous) injection

$$\mathscr{H}_0 \xrightarrow{J} E$$

and its unique extension to

$$\hat{\mathscr{H}}_0 \xrightarrow{\hat{J}} E\,(= \hat{E}).$$

We must show that $\hat{J}$ is still an injection. Let $h \in \hat{\mathscr{H}}_0$ and let $\dot{h} = \hat{J}h \in E$. We claim that for every $e' \in E'$

$$\langle e', \dot{h}\rangle = (h|Le')_{\hat{\mathscr{H}}_0} \qquad 2.4)$$

If $h \in \mathscr{H}_0$, then $h = \dot{h} = L\dot{f}'$ and equation (2.4) simply reduces to the definition (2.3) of the scalar product in $\mathscr{H}_0$. Consider now the sequence $\{h_\nu\} \subset \mathscr{H}_0$ such that $h_\nu \to h$. Then

$$\langle e', h_\nu\rangle = (h_\nu|Le')_{\mathscr{H}_0}.$$

Passing to the limit and using the continuity of the scalar product in $\hat{\mathscr{H}}_0$ and the continuity of the form e' we obtain equation (2.4). Suppose now that $\dot{h} = \hat{J}h = 0$. Then, by equation (2.4), $(h|Le')_{\hat{\mathscr{H}}_0} = 0$ for all $Le' \in \mathscr{H}_0$, and since $\mathscr{H}_0$ is dense in $\hat{\mathscr{H}}_0$ it follows $h = 0$. Thus $\hat{J}$ is one-to-one.

(3) Finally let $\mathscr{H} = \hat{J}\hat{\mathscr{H}}_0$ and transfer the Hilbert structure of $\hat{\mathscr{H}}_0$ to $\mathscr{H}$. We must show that $\mathscr{H}$ is associated with L. If $k \in \mathscr{H}$, then there is an $h \in \hat{\mathscr{H}}_0$ such that $\hat{J}h = \dot{h} = k$ and equation (2.4) yields

$$\langle e', k\rangle = \langle e', \dot{h}\rangle = (h|Le')_{\hat{\mathscr{H}}_0} = (k|Le')_{\mathscr{H}}$$

which shows that L is the anti-kernel associated with $\mathscr{H}$.

We shall give now another construction of the Hilbert space $\mathscr{H}$ corresponding to a given positive anti-kernel L. This construction will be from above, in contrast to the one given in the proof of the previous theorem, which was from below.

Theorem. Let L be a positive anti-kernel. An element $h \in E$ belongs to the Hilbert space $\mathscr{H}$ corresponding to L if and only if

$$\sup_{e' \in E'} \frac{|\langle e', h \rangle|}{\langle e', Le' \rangle^{\frac{1}{2}}} < \infty, \tag{2.4}$$

and if this condition is satisfied, then

$$\|h\|_{\mathscr{H}} = \sup_{e' \in E'} \frac{|\langle e', h \rangle|}{\langle e', Le' \rangle^{\frac{1}{2}}}.$$

Proof. If $h \in \mathscr{H}$, then, using equation (2.1) and Schwarz's inequality,

$$|\langle e', h \rangle| = |(h|Le')_{\mathscr{H}}| \leqslant (h|h)^{\frac{1}{2}}_{\mathscr{H}} (Le'|Le')^{\frac{1}{2}}_{\mathscr{H}} = \|h\|_{\mathscr{H}} \langle e', Le' \rangle^{\frac{1}{2}}$$

for all $e' \in E'$.

Conversely, suppose that equation (2.4) holds and consider the map

$$Le' \to \langle e', h \rangle$$

It may be easily verified that this map is an anti-linear functional on $\mathscr{H}_0$ which is continuous, since it is bounded on the unit ball of $\mathscr{H}_0$. Therefore, it may be continued to a continuous anti-linear functional on the completion $\mathscr{H}$ of $\mathscr{H}_0$. Hence, by the Riesz representation theorem there exists $k \in \mathscr{H}$ such that

$$Le' \to \langle e', h \rangle = (k|Le')_{\mathscr{H}} = \langle e', k \rangle$$

for all $e' \in E'$. By Hahn-Banach theorem, $h = k \in \mathscr{H}$.

The Structures of $\mathfrak{H}$ and $\overline{\mathscr{L}}_+ (E', E)$

Let $\mathscr{L}(E', E)$ denote the set of continuous linear maps from E' into E, $\overline{\mathscr{L}}(E', E)$ denote the set of continuous anti-linear maps, or anti-kernels, from E' into E and $\overline{\mathscr{L}}_+(E', E)$ denote the set of positive anti-kernels. It was shown in the previous section that there is a one-to-one correspondence between $\mathfrak{H}$ and $\overline{\mathscr{L}}_+(E', E)$, $\mathfrak{H} \approx \overline{\mathscr{L}}_+(E', E)$. It was also mentioned that we may define a natural structure (addition, scalar multiplication, etc.) in both $\mathfrak{H}$ and $\overline{\mathscr{L}}_+(E', E)$. In this section the structures of $\mathfrak{H}$ and $\overline{\mathscr{L}}_+(E', E)$ will be defined more precisely and the correspondence between them will be established.

Definition. In $\overline{\mathscr{L}}_+ (E', E)$ we define the following:

(1) *Order relation*: $L_1 \leqslant L_2$ if $L_2 - L_1 \geqslant 0$ i.e. if $L_2 - L_1$ is a positive anti-kernel.

(2) *Multiplication by a non-negative scalar* λ: $(\lambda L)(e') = L(\lambda e')$.

(3) *Addition*: $(L_1 + L_2)e' = L_1e' + L_2e'$.

Correspondingly in $\mathfrak{H}$ we define the following:

(1′) *Order relation*: $\mathscr{H}_1 \leqslant \mathscr{H}_2$ if $\mathscr{H}_1 \subset \mathscr{H}_2$ and the norm in $\mathscr{H}_1$ is greater than or equal to the norm in $\mathscr{H}_2$, $\|h\|_{\mathscr{H}_1} \geqslant \|h\|_{\mathscr{H}_2}$.

(2′) *Multiplication by a non-negative scalar* λ:

To $\mathscr{H}$ corresponds the space

$$\lambda\mathscr{H} = \begin{cases} \{0\} \text{ if } \lambda = 0 \\ \mathscr{H} \text{ if } \lambda > 0, \end{cases}$$

and the norm

$$\|h\|_{\lambda\mathscr{H}} = 1/\sqrt{(\lambda)}\,\|h\|_{\mathscr{H}}.$$

(3′) *Addition*: To $(\mathscr{H}_1, \mathscr{H}_2)$ corresponds the space

$$\mathscr{H}_1 + \mathscr{H}_2 = \{h : h \in E, h = h_1 + h_2, h_1 \in \mathscr{H}_1, h_2 \in \mathscr{H}_2\}$$

and the norm

$$\|h\|_{\mathscr{H}_1 + \mathscr{H}_2} = \inf \sqrt{(\|h_1\|^2_{\mathscr{H}_1} + \|h_2\|^2_{\mathscr{H}_2})}$$

It should be noted that (1) and (1′) indeed define order relations since we have:

(a) $\mathscr{H}_1 \leqslant \mathscr{H}_2$ and $\mathscr{H}_2 \leqslant \mathscr{H}_1$ imply that $\mathscr{H}_1 = \mathscr{H}_2$.

(b) $L_1 \leqslant L_2$ and $L_2 \leqslant L_1$ imply that $L_1 = L_2$.

The first follows trivially from the fact that, in a Hilbert space, the scalar product is uniquely determined by the norm. The hypothesis of the second means, by definition, that

$$\langle e', (L_2 - L_1)e' \rangle \geqslant 0, \langle e', (L_1 - L_2)e' \rangle \geqslant 0, e' \in E',$$

or

$$\langle e', L_1e' \rangle = \langle e', L_2e' \rangle, e' \in E'.$$

Using now the formula

$$4\langle e', Lf' \rangle = \langle e' + f', L(e' + f') \rangle - \langle e' - f', L(e' - f') \rangle$$
$$+ i\langle e' + if', L(e' + if') \rangle - i\langle e' - if', L(e' - if') \rangle$$

we conclude that

$$\langle e', L_1 f' \rangle = \langle e', L_2 f' \rangle, \text{ for every } e', f' \in E'$$

and by Hahn-Banach theorem it follows that

$$L_1 f' = L_2 f', \qquad f' \in E'.$$

It should also be mentioned, as a consequence of the requirement $\|h\|_{\mathscr{H}_1} \geqslant \|h\|_{\mathscr{H}_2}$ in the definition of $\mathscr{H}_1 \leqslant \mathscr{H}_2$, that convergence in $\mathscr{H}_1$ implies convergence in $\mathscr{H}_2$.

We shall prove now the correspondence between the structures of $\mathfrak{H}$ and $\overline{\mathscr{L}}_+(E', E)$.

Theorem. (1) If $\mathscr{H}_1$ and $\mathscr{H}_2$ correspond to L_1 and L_2, respectively, then $\mathscr{H}_1 \leqslant \mathscr{H}_2$ if and only if $L_1 \leqslant L_2$.
(2) If $\mathscr{H}$ corresponds to L, then $\lambda\mathscr{H}$ corresponds to $\lambda L (\lambda \geqslant 0)$.
(3) If $\mathscr{H}_1$ and $\mathscr{H}_2$ correspond to L_1 and L_2, respectively, then $\mathscr{H}_1 + \mathscr{H}_2$ corresponds to $L_1 + L_2$.

Proof. (1) Suppose first that $\mathscr{H}_1 \leqslant \mathscr{H}_2$. Then

$$L_1 e' \in \mathscr{H}_1 \subset \mathscr{H}_2 \text{ and}$$

$$\langle e', L_1 e'\rangle^{\frac{1}{2}} = \|L_1 e'\|_{\mathscr{H}_1} \geqslant \|L_1 e'\|_{\mathscr{H}_2} \geqslant \frac{|\langle f', L_1 e'\rangle|}{\langle f', L_2 f'\rangle^{\frac{1}{2}}}, \; e', f' \in E'$$

choosing $f' = e'$,

$$\langle e', L_1 e'\rangle^{\frac{1}{2}} \geqslant \frac{\langle e', L_1 e'\rangle}{\langle e', L_2 e'\rangle^{\frac{1}{2}}},$$

or

$$\langle e', L_2 e'\rangle^{\frac{1}{2}} \geqslant \langle e', L_1 e'\rangle^{\frac{1}{2}}$$

or

$$\langle e', (L_2 - L_1) e'\rangle \geqslant 0$$

or

$$L_1 \leqslant L_2.$$

Conversely, suppose that $L_1 \leqslant L_2$. Then, for any $h \in E$:

$$\frac{|\langle e', h\rangle|}{\langle e', L_1 e'\rangle^{\frac{1}{2}}} \geqslant \frac{|\langle e', h\rangle|}{\langle e', L_2 e'\rangle^{\frac{1}{2}}};$$

hence $\mathscr{H}_1 \subset \mathscr{H}_2$ and, if $h \in \mathscr{H}_1$, $\|h\|_{\mathscr{H}_1} \geqslant \|h\|_{\mathscr{H}_2}$.
(2) This follows immediately from the choice of the norm in $\lambda\mathscr{H}$:

$$\|h\|_{\lambda\mathscr{H}} = \frac{1}{\sqrt{\lambda}} \sup \frac{|\langle e', h\rangle|}{\langle e', L e'\rangle^{\frac{1}{2}}} = \sup \frac{|\langle e', h\rangle|}{\langle e', \lambda L e'\rangle^{\frac{1}{2}}}$$

(3) We must first say what we mean by the Hilbert space $\mathscr{H}_1 + \mathscr{H}_2$.

The *set* $\mathscr{H}_1 + \mathscr{H}_2$ consists of all elements h of E which may be written in the form $h = h_1 + h_2$ with $h_1 \in \mathscr{H}_1$ and $h_2 \in \mathscr{H}_2$. If $\mathscr{H}_1 \cap \mathscr{H}_2 = \{0\}$ the norm in $\mathscr{H}_1 + \mathscr{H}_2$ may be defined by

$$\|h\|^2_{\mathscr{H}_1+\mathscr{H}_2} = \|h_1\|^2_{\mathscr{H}_1} + \|h\|^2_{\mathscr{H}_2}$$

where $h = h_1 + h_2$ is the unique representation of h. If, however, $\mathscr{H}_1 \cap \mathscr{H}_2 \neq \{0\}$, then an element $h \in \mathscr{H}_1 + \mathscr{H}_2$ has an infinite number of representations since the zero element has an infinite number of representations of the form $a - a$, where $a \in \mathscr{H}_1 \cap \mathscr{H}_2$. For this reason we take as the definition of the norm

$$\|h\|^2_{\mathscr{H}_1+\mathscr{H}_2} = \inf(\|h_1\|^2_{\mathscr{H}_1} + \|h_2\|^2_{\mathscr{H}_2})$$

where the infimum is taken over all possible representations $h = h_1 + h_2$. It must be shown that this is actually a norm, that it may be defined by a scalar product in $\mathscr{H}_1 + \mathscr{H}_2$, that $\mathscr{H}_1 + \mathscr{H}_2$ is complete and the topology of $\mathscr{H}_1 + \mathscr{H}_2$ is finer than the topology of E. Since this procedure seems rather tedious, we shall construct the Hilbert space $\mathscr{H}_1 + \mathscr{H}_2$ in a different way.

Let $\mathscr{H}_1 \oplus \mathscr{H}_2$ denote the abstract Hilbertian (direct) sum of $\mathscr{H}_1$ and $\mathscr{H}_2$. An element of $\mathscr{H}_1 \oplus \mathscr{H}_2$ is a pair (h_1, h_2) where $h_1 \in \mathscr{H}_1$ and $h_2 \in \mathscr{H}_2$. Addition and scalar multiplication are defined by

$$(h_1, h_2) + (k_1, k_2) = (h_1 + k_1, h_2 + k_2),$$
$$\lambda(h_1, h_2) = (\lambda h_1, \lambda h_2),$$

and the scalar product in $\mathscr{H}_1 \oplus \mathscr{H}_2$ is defined by

$$((h_1, h_2)|(k_1, k_2))_{\mathscr{H}_1 \oplus \mathscr{H}_2} = (h_1|k_1)_{\mathscr{H}_1} + (h_2|k_2)_{\mathscr{H}_2}$$

Note that $\mathscr{H}_1 \approx \mathscr{H}_1 \oplus \{0\}$ and $\mathscr{H}_2 \approx \{0\} \oplus \mathscr{H}_2$. Let us map the abstract Hilbert space $\mathscr{H}_1 \oplus \mathscr{H}_2$ into E by

$$(h_1, h_2) \to h_1 + h_2 \in E.$$

Let $\mathscr{N} = \{(h_1, h_2): h_1 + h_2 = \{0\}\}$ be the null space of this map. Clearly, $\mathscr{N}$ is closed in $\mathscr{H}_1 \oplus \mathscr{H}_2$. Hence the factor space $\mathscr{H}_1 \oplus \mathscr{H}_2/\mathscr{N}$ is canonically a Hilbert space which can be identified with the orthogonal complement of $\mathscr{N}$ in $\mathscr{H}_1 \oplus \mathscr{H}_2$. Consider now the maps

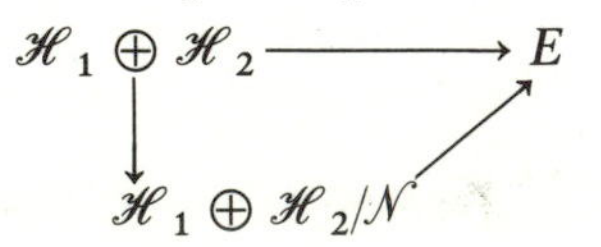

The map $\mathscr{H}_1 \oplus \mathscr{H}_2/\mathscr{N} \to E$ is an injection whose image is the space $\mathscr{H}_1 + \mathscr{H}_2$. We can transfer the structure of $\mathscr{H}_1 \oplus \mathscr{H}_2/\mathscr{N}$ on $\mathscr{H} + \mathscr{H}_2$; it is seen immediately that the factor norm on $\mathscr{H}_1 \oplus \mathscr{H}_2/\mathscr{N}$ is the infimum norm defined above in $\mathscr{H}_1 + \mathscr{H}_2$. By this construction, the space $\mathscr{H}_1 + \mathscr{H}_2$ is already proved to be a Hilbert space contained in E, whose topology is finer than that of E.

We must still show that $\mathscr{H}_1 + \mathscr{H}_2$ corresponds to $L_1 + L_2$. Let $L = L_1 + L_2$ and let $\mathscr{K}$ denote the Hilbert space corresponding to L. We must show that $\mathscr{K} = \mathscr{H}_1 + \mathscr{H}_2$. Since $L \geqslant L_1$ and $L \geqslant L_2$ it follows that $\mathscr{K} \supset \mathscr{H}_1$ and $\mathscr{K} \supset \mathscr{H}_2$; hence $\mathscr{K} \supset \mathscr{H}_1 + \mathscr{H}_2$. Furthermore,

$$\mathscr{K}_0 = \{Le' : e' \in E'\}$$

is contained in $\mathscr{H}_1 + \mathscr{H}_2$ since $Le' = L_1 e' + L_2 e'$. Thus we have

$$\mathscr{K} \supset \mathscr{H}_1 + \mathscr{H}_2 \supset \mathscr{K}_0$$

where $\mathscr{K}_0$ is a dense subset of $\mathscr{K}$. Since both $\mathscr{K}$ and $\mathscr{H}_1 + \mathscr{H}_2$ are complete Hilbert spaces it suffices to show that the norm of $\mathscr{K}_0$ is equal to the norm of $\mathscr{H}_1 + \mathscr{H}_2$. Let $h = Le' = L_1 e' + L_2 e' = h_1 + h_2$, where $h_1 = L_1 e' \in \mathscr{H}_1$ and $h_2 = L_2 e' \in \mathscr{H}_2$. We have

$$\langle e', Le' \rangle = \langle e', L_1 e' \rangle + \langle e', L_2 e' \rangle$$

or

$$\|h\|^2_{\mathscr{K}_0} = \|h_1\|^2_{\mathscr{H}1} + \|h_2\|^2_{\mathscr{H}2} \geqslant \|h\|^2_{\mathscr{H}_1 + \mathscr{H}_2}$$

In order that equality holds it is sufficient to show that the element (h_1, h_2) of $\mathscr{H}_1 \oplus \mathscr{H}_2$ is orthogonal to the null space $\mathscr{N}$, because in this case there is no reduction in norm. If $(n_1, n_2) \in \mathscr{N}$ we have

$$\begin{aligned}(h_1|n_1)_{\mathscr{H}_1} + (h_2|n_2)_{\mathscr{H}_2} &= (L_1 e'|n_1)_{\mathscr{H}_1} + (L_2 e'|n_2)_{\mathscr{H}_2} \\ &= \overline{\langle e', n_1 \rangle} + \overline{\langle e', n_2 \rangle} = \overline{\langle e', n_1 + n_2 \rangle} = \overline{\langle e', 0 \rangle} = 0.\end{aligned}$$

Remark. If $\mathscr{H}_1 \leqslant \mathscr{H}$, then there exists one and only one Hilbert space $\mathscr{H}_2$ such that $\mathscr{H}_1 + \mathscr{H}_2 = \mathscr{H}$. This follows from the fact that $L_2 = L - L_1$, is a positive anti-kernel.

Definition. The anti-kernels L_1 and L_2 are called *disjoint* if the only anti-kernel which is less than or equal to both L_1 and L_2 is zero, i.e., if $L \leqslant L_1$ and $L \leqslant L_2$ implies $L = 0$.

Definition. The convergence of a sequence of anti-kernels $\{L_j\} \subset \overline{\mathscr{L}}_+(E', E)$ to an anti-kernel L is called

(1) *point-wise convergence* if $L_j e' \to Le'$ for all $e' \in E'$.

(2) *bounded convergence* if $L_j e' \to Le'$ for all $e' \in E'$ and this convergence is uniform on bounded subsets of E'.

It should be noted that $\overline{\mathscr{L}}_+(E', E)$ is a closed subset of $\overline{\mathscr{L}}(E', E)$ under both of the above convergences. Furthermore it is a convex cone. It is a cone with vertex the origin since tL is a positive anti-kernel for $t \geqslant 0$ provided that L is positive, and it is convex since $tL_1 + (1 - t)L_2$ is a positive anti-kernel for every $0 \leqslant t \leqslant 1$ provided that L_1 and L_2 are positive. We may therefore state the following.

Theorem. There is a one-to-one correspondence between the space $\mathfrak{H}$ and a closed convex cone in a topological vector space. This cone does not contain any vector subspace other than $\{0\}$. The last sentence of this statement expresses the fact that the relation $\mathscr{H}_1 \leqslant \mathscr{H}_2$ is indeed an order relation.

We turn now to the following question. Given a system $\{\theta_i\}$ of elements of E, what are the necessary and sufficient conditions in order that it forms a complete orthonormal system for a Hilbert space $\mathscr{H} \in \mathfrak{H}$? We observe first that an element $e \in E$ defines the following maps:

(a) a linear functional on E':

$$f' \to \langle f', e \rangle, \qquad f' \in E',$$

(b) an anti-linear functional on E':

$$f' \to \overline{\langle f', e \rangle}, \qquad f' \in E'$$

(c) a hermitian form on $E' \times E'$:

$$(f', g') \to \langle f', e \rangle \overline{\langle g', e \rangle}$$

(d) a map from E' into E, denoted by $e\bar{e}$:

$$f' \xrightarrow{e\bar{e}} \overline{\langle f', e \rangle}\, e$$

It can be easily shown that $e\bar{e}$ is a positive anti-kernel. The corresponding Hilbert space $\mathscr{H} = \{\lambda e : \lambda \in C\}$ with the norm $\|\lambda e\| = |\lambda|$ and the scalar product $(\lambda e | \mu e) = \lambda \bar{\mu}$.

Theorem. In order that a given set $\{e_i\}_{i \in I}$ of elements of E be a Hilbert basis, i.e., a complete orthonormal system of a Hilbert space $\mathscr{H} \in \mathfrak{H}$, it is necessary and sufficient that:

(1) The series of anti-kernels $\sum_{i \in I} e_i \bar{e}_i$ is pointwise convergent or that the finite partial sums of the series are pointwise bounded;

(2) The set $\{e_i\}_{i\in I}$ is Hilbert-free, i.e. if $\{c_i\}_{i\in I}$ is any set of complex numbers such that $\sum_{i\in I} |c_i|^2 < \infty$ and if $\sum c_i e_i = 0$, then $c_i = 0$ for all $i \in I$.

Proof. Suppose that $\{e_i\}_{i\in I}$ is a Hilbert basis of $\mathscr{H} \in \mathfrak{H}$ and let L be the corresponding anti-kernel. Then, for every $f' \in E'$,

$$\sum_{i\in I} e_i\, \bar{e}_i\, .f' = \sum_{i\in I} e_i\, \overline{\langle f', e_i\rangle} = \sum_{i\in I} e_i (Lf' | e_i) \quad = Lf'$$

so that $\sum_{i\in I} e_i \bar{e}_i$ converges pointwise.

Furthermore, since $\{e_i\}_{i\in I}$ is a Hilbert basis, it follows immediately that it is Hilbert-free.

Conversely, assuming that (1) and (2) hold, let us show that $\{e_i\}_{i\in I}$ is a Hilbert basis of some $\mathscr{H} \in \mathfrak{H}$. We first notice that condition (1) means that for every $f' \in E'$ the partial sums of $\sum_{i\in I} \overline{\langle e_i, f'\rangle}\, e_i$ are strongly bounded in E, and for every f' and g' in E', the partial sums of $\sum_{i\in I} \overline{\langle e_i, f'\rangle} \langle e_i, g'\rangle$ are bounded. Taking $f' = g'$ and using an elementary result about series of numbers $\geqslant 0$, we conclude that

$$\sum_{i\in I} |\langle e_i, f'\rangle|^2 < \infty, f' \in E'.$$

We shall attempt now to construct the Hilbert space $\mathscr{H}$. Consider the Hilbert space l^2 whose elements are families $\{x_i\}_{i\in I}$ of complex numbers such that $\sum_{i\in I} |x_i|^2 < \infty$, and let l_0^2 denote the subset of l^2 whose elements have all but a finite number of co-ordinates equal to zero. There is a natural map from l_0^2 to E given by

$$\{x_i\}_{i\in I} \to \sum_{i\in I} x_i\, e_i.$$

In order to show that this map is continuous, it is sufficient to show that the image of the unit ball is bounded, or, by Mackey's theorem, that it is weakly bounded. This follows from Schwarz's inequality. Let $\{x_i\}_{i\in I} \in l_0^2$ such that $\sum_{i\in I} |x_i|^2 \leqslant 1$. Then, for every $f' \in E'$,

$$|\langle \sum x_i e_i, f'\rangle| = |\sum x_i \langle e_i, f'\rangle| \leqslant$$

$$\leqslant (\sum |x_i|^2)^{\frac{1}{2}} (\sum |\langle e_i, f'\rangle|^2)^{\frac{1}{2}} \leqslant$$

$$\leqslant (\sum |\langle e_i, f'\rangle|^2)^{\frac{1}{2}} < \infty.$$

Let us extend now the above map to the completion of l_0^2, which can be identified with l^2:

$$l^2 \to E$$

by

$$\{x_i\}_{i \in I} \to \sum_{i \in I} x_i \bar{e}_i.$$

Using the Hilbert-free hypothesis (2) it may be shown that this map is an injection. Let $\mathscr{H}$ be the image of this injection with structure the transferred structure from l^2. The set $\{e_i\}_{i \in I}$ is the image of the canonical basis of l^2 and hence it is a Hilbert basis for $\mathscr{H}$.

Corollary. If L is a positive anti-kernel, then L has an infinity of decompositions of the form

$$L = \sum e_i \bar{e}_i.$$

Scalar Particles

Let us return now to the study of scalar particles in a universe V. Recall that a universe is a C^∞-manifold of dimension n and a scalar particle is a Hilbert space $\mathscr{H}$ with continuous injection in the space $\mathscr{D}'(V)$ of distributions on the universe V. The locally convex topological vector space $\mathscr{D}'(V)$ is the dual of the space $\mathscr{D}(V)$ of infinitely differentiable functions with compact support on V. An element T of $\mathscr{D}'(V)$ is a distribution whose value on an element φ of $\mathscr{D}(V)$ is denoted by $\langle T, \varphi \rangle$; $\mathscr{D}(V)$ is a reflexive space, i.e., the dual of $\mathscr{D}'(V)$ is $\mathscr{D}(V)$. Each of the spaces $\mathscr{D}(V)$ and $\mathscr{D}'(V)$ has the strong dual topology of the other. We shall apply now the results of the last two sections for $E = \mathscr{D}'(V)$ and $E' = \mathscr{D}(V)$.

In order to find the Hilbert spaces in $\mathscr{D}'(V)$ we must look for the anti-kernels $L\colon \mathscr{D}(V) \to \mathscr{D}'(V)$. We shall start by looking for the continuous linear maps from $\mathscr{D}(V)$ into $\mathscr{D}'(W)$ where V and W are two C^∞-manifolds, for example, two Euclidean spaces. Let y denote a generic point of V and x denote a generic point of W. For convenience, we may denote $\mathscr{D}(V)$ by $\mathscr{D}_y$ and $\mathscr{D}'(W)$ by $\mathscr{D}'_x$. Let $\mathscr{D}'(W \times V)$ or $\mathscr{D}'_{x,y}$ denote the space of distributions on the product $W \times V$ (distributions of two variables).

Theorem. (Theorem of kernels). The topological vector space $\mathscr{L}_b$ $(\mathscr{D}(V); \mathscr{D}'(W))$ of continuous linear maps from $\mathscr{D}(V)$ into $\mathscr{D}'(W)$ with the topology of bounded convergence is canonically isomorphic with the topological vector space $\mathscr{D}'(W \times V)$.

Let K be a given element of $\mathscr{D}'(W \times V)$. K defines a continuous linear map $v \to K \cdot v$ from $\mathscr{D}(V)$ into $\mathscr{D}'(W)$ by the formula

$$\langle K \cdot v, w \rangle = \langle K, w \otimes v \rangle, \ w \in \mathscr{D}(W) \tag{2.5}$$

where $w \otimes v = w(x) \cdot v(y)$. We must verify first that $K \cdot v \in \mathscr{D}'(W)$. Clearly $K \cdot v$ is a linear functional in w. If $w \to 0$ in $\mathscr{D}(W)$, then $w \otimes v \to 0$ in $\mathscr{D}(W \times V)$, and, since $K \in \mathscr{D}'(W \times V)$, $\langle K \cdot v, w \rangle = \langle K, w \otimes v \rangle \to 0$. We must also show that $v \to K \cdot v$ is a continuous linear map. The linearity is obvious. If $v \to 0$ in $\mathscr{D}(V)$, then it is easily seen that $\langle K \cdot v, w \rangle \to 0$ for every $w \in \mathscr{D}(W)$, and this convergence is uniform when w remains bounded in $\mathscr{D}(W)$.

For more clarity we shall repeat this result in terms of the alternate notation $\mathscr{D}_y$, $\mathscr{D}'_x$, $\mathscr{D}_{x,y}$, and we shall give an example from the theory of integral equations from which this notation was originated. Let $K_{x,y} \in \mathscr{D}'_{x,y}$ be a given distribution in the two variables x and y. $K_{x,y}$ defines a continuous linear map from $\mathscr{D}_y$ into $\mathscr{D}'_x$:

$$v(y) \to K_{x,y} \cdot v(y) = (K \cdot v)_x, \quad v(y) \in \mathscr{D}_y,$$

where $(K \cdot v)(x) \in \mathscr{D}'_x$ is defined by the formula

$$\langle (K \cdot v)_x, w(x) \rangle = \langle K_{x,y}\, w(x) v(y) \rangle, \quad w(x) \in \mathscr{D}_x.$$

Example. Let $V_{(y)}$ and $W_{(x)}$ be two Euclidean spaces equipped with a Lebesgue measure, and let $K(x, y)$ be a locally integrable function on $W \times V$. $K(x, y) \in \mathscr{D}'_{x,y}$ is a distribution in the two variables x and y. Let

$$\langle T, \varphi \rangle = \int T\varphi$$

be the value of a distribution T at the testing function φ. $K(x, y)$ defines a continuous linear map from $\mathscr{D}_y$ into $\mathscr{D}'_x$:

$$K : v(y) \to (K \cdot v)_x, \quad v \in \mathscr{D}_y$$

where $(K \cdot v)_x \in \mathscr{D}'_x$ is defined in accordance with our formula

$$\langle K, w \otimes v \rangle = \iint K(x, y) w(x) v(y)\, dx\, dy$$
$$= \int \left[\int K(x, y) v(y)\, dy \right] w(x)\, dx = \langle K \cdot v, w \rangle,$$

where the Fubini theorem was used. Thus

$$(K \cdot v)_x = \int K(x, y) v(y)\, dy.$$

The converse part of the theorem of kernels asserts that every continuous linear map from $\mathscr{D}(V)$ into $\mathscr{D}'(W)$ defines a unique distribution on $W \times V$. The proof of this is considerably more difficult, and it will be omitted. Instead, we shall turn now to the proof of the equivalence of topologies of the isomorphic spaces $\mathscr{L}(\mathscr{D}_y, \mathscr{D}'_x)$ and $\mathscr{D}'_{x,y}$. There is a natural topology in both these spaces. In $\mathscr{D}'_{x,y}$ we shall take, as previously, the strong topology (uniform convergence on bounded subsets of $\mathscr{D}_{x,y}$). In $\mathscr{L}(\mathscr{D}_y, \mathscr{D}'_x)$ we have point-wise and bounded convergence. Let $\mathscr{L}_b(\mathscr{D}_y, \mathscr{D}'_x)$ denote the topological vector space consisting of the vector space $\mathscr{L}(\mathscr{D}_y, \mathscr{D}'_x)$ with the topology of bounded convergence. A sequence in $\mathscr{L}_b(\mathscr{D}_y, \mathscr{D}'_x)$ of continuous linear maps from $\mathscr{D}_y$ into $\mathscr{D}'_x$ converges to zero in the sense of bounded convergence if the image sequence of every element in $\mathscr{D}_y$ converges to zero in $\mathscr{D}'_x$, and this convergence is uniform on bounded subsets of $\mathscr{D}_y$. We shall prove only that convergence in $\mathscr{D}'_{x,y}$ implies convergence in $\mathscr{L}_b(\mathscr{D}_y, \mathscr{D}'_x)$. Let $\{K_j\}$ be a sequence in $\mathscr{D}'_{x,y}$ such that $K_j \to 0$ strongly, i.e., $\langle K_j, \varphi(x, y)\rangle \to 0$ for every $\varphi(x, y) \in \mathscr{D}_{x,y}$, and this convergence is uniform on bounded subsets of $\mathscr{D}_{x,y}$. We must show that for a fixed $v(y) \in \mathscr{D}_y$, $\langle K_j \cdot v, w\rangle \to 0$ for every $w(x) \in \mathscr{D}_x$, and this convergence is uniform when $w(x)$ runs over bounded subsets of $\mathscr{D}_x$. By definition,

$$\langle K_j \cdot v, w\rangle = \langle K_j(x, y), w(x)v(y)\rangle \to 0$$

and the last convergence is uniform when $w(x)$ remains bounded, since in this case, $w(x)v(y)$ remains bounded. Again, the proof of the converse will be omitted.

Remark. The Banach–Steinhaus theorem implies that any weakly convergent sequence in $\mathscr{D}'$ is also strongly convergent.

Now that we have identified the space $\mathscr{L}_b(\mathscr{D}(V); \mathscr{D}'(V))$ of continuous linear maps from $\mathscr{D}(V)$ into $\mathscr{D}'(V)$ with the space $\mathscr{D}'(V \times V)$ of distributions of two variables, it is a simple matter to obtain the space $\overline{\mathscr{L}}_+(\mathscr{D}(V); \mathscr{D}'(V))$ of positive anti-kernels from $\mathscr{D}(V)$ to $\mathscr{D}'(V)$ and the space $\mathfrak{H}$ of Hilbert spaces $\mathscr{H}$ with continuous injection in $\mathscr{D}'(V)$. A distribution of two variables $K_{x,y} \in \mathscr{D}'(V \times V)$ defines a continuous linear map from $\mathscr{D}(V)$ into $\mathscr{D}'(V)$:

$$v \to K \cdot v.$$

Since we want an anti-linear map, we shall take, instead,

$$v \to K \cdot \overline{v}$$

where $\overline{v}$ is the complex conjugate of the scalar-valued function $v \in \mathscr{D}(V)$.

The positivity of this anti-kernel is defined naturally as follows:

Definition. The anti-kernel $v \to K \cdot \bar{v}$ defined by the distribution $K_{x,y} \in \mathscr{D}'(V \times V)$ is called *positive* if

$$\langle K_{x,y}, \varphi(x) \otimes \overline{\varphi(y)} \rangle \geqslant 0$$

for every $\varphi \in \mathscr{D}(V)$.

According to our general theorem, the relation between the positive E-anti-kernel L and the corresponding Hilbert space $\mathscr{H}$ is

$$\langle h, e' \rangle = (h | Le')_{\mathscr{H}},$$

for every $h \in \mathscr{H}$ and every $e' \in E'$. In our case, the relation between the positive anti-kernel K from $\mathscr{D}(V)$ into $\mathscr{D}'(V)$ with the corresponding $\mathscr{H}$ is

$$\langle T, \varphi \rangle = (T | K \cdot \bar{\varphi})_{\mathscr{H}} \tag{2.6}$$

for every distribution $T \in \mathscr{H} \subset \mathscr{D}'(V)$ and every $\varphi \in \mathscr{D}(V)$. If we let $T = K \cdot \bar{\psi}$ in the above formula, we obtain

$$\langle K \cdot \bar{\psi}, \varphi \rangle = \langle K, \varphi \otimes \bar{\psi} \rangle = (K \cdot \bar{\psi} | K \cdot \bar{\varphi}) \tag{2.7}$$

for every φ and ψ in $\mathscr{D}(V)$.

Remarks. In formula (2.6) no error should be made by putting a bar over T. The space $\mathscr{H}$ is not invariant under conjugation. The space $\overline{\mathscr{H}} = \{\bar{T}: T \in \mathscr{H}\}$ represents the same particle as $\mathscr{H}$ but with opposite charge.

The definition of positivity of the anti-kernel defined by $K_{x,y}$ originated from the theory of integral equations. Let $K(x, y)$ be a continuous function on $V \times V$. The kernel $K(x, y)$ is called positive if for every set of elements $\{x_1, x_2, \ldots, x_l\}$ in the space V and every set of complex numbers $\{z_1, z_2, \ldots, z_l\}$ the following inequality holds:

$$\sum_{i,j}^{l} K(x_i, x_j) z_i \bar{z}_j \geqslant 0.$$

It may be shown that the definition of positivity of the anti-kernel defined by $K_{x,y} \in \mathscr{D}'(V \times V)$ coincides with the above definition when $K_{x,y}$ is a continuous function $K(x, y)$ of the two variables x and y.

We summarize the final results of this section in the following theorem:

Theorem. Let V be a C^∞-manifold. The space $\mathfrak{H}$ of the Hilbert spaces $\mathscr{H}$ with continuous injection in $\mathscr{D}'(V)$ in canonically isomorphic to the subspace of $\mathscr{D}'(V \times V)$ consisting of distributions $K_{x,y}$ of two variables such that

$$\langle K, \varphi \otimes \overline{\varphi} \rangle \geqslant 0$$

for every $\varphi \in \mathscr{D}(V)$. The relation between such a K and the corresponding $\mathscr{H}$ is given by the formula

$$\langle T, \varphi \rangle = (T | K \cdot \overline{\varphi})_{\mathscr{H}}, \quad T \in \mathscr{H}, \quad \varphi \in \mathscr{D}(V),$$

where the distribution $K \cdot \overline{\varphi} \in \mathscr{H} \subset \mathscr{D}'(V)$ is defined by

$$\langle K \cdot \overline{\varphi}, \psi \rangle = \langle K, \psi \otimes \overline{\varphi} \rangle, \quad \psi \in \mathscr{D}(V).$$

Tensor Products

In order to generalize the results of the previous section to vector particles we must introduce the concept of a tensor product. We shall present only the basic definition and properties without any proofs.

Let E and F be two vector spaces. For our purpose, it is not necessary to give a complete definition of the tensor product. The tensor product of E and F is a new vector space $E \otimes F$ with a given canonical bi-linear map from $E \times F$ into $E \otimes F$:

$$\mathbf{e} \times \mathbf{f} \rightarrow \mathbf{e} \otimes \mathbf{f}$$

$E \otimes F$ is not the image of $E \times F$ under this map. However, $E \otimes F$ is generated by elements of the form $\mathbf{e} \otimes \mathbf{f}$, i.e., every element $\mathbf{x}$ of $E \otimes F$ may be written as

$$\mathbf{x} = \mathbf{e}_1 \otimes \mathbf{f}_1 + \mathbf{e}_2 \otimes \mathbf{f}_2 + \ldots + \mathbf{e}_k \otimes \mathbf{f}_k.$$

The image under the canonical bi-linear map is not a vector subspace of $E \otimes F$, but $E \otimes F$ is formed by finite sums of elements of the form $\mathbf{e} \otimes \mathbf{f}$. We state now the following properties.

If E and F have finite dimensions m and n, respectively, then the dimension of $E \otimes F$ is $m \cdot n$.

If $\{\mathbf{e}_i\}$ and $\{\mathbf{f}_j\}$ are bases of E and F, respectively, then $\{\mathbf{e}_i \otimes \mathbf{f}_j\}$ is a basis of $E \otimes F$, i.e., every element $\mathbf{x}$ of $E \otimes F$ may be written in a unique way as

$$\mathbf{x} = \sum_{i,j} x_{ij} \mathbf{e}_i \otimes \mathbf{f}_j$$

If F is finite-dimensional with $\{f_i\}$ as its basis, then every element

$\mathbf{x} \in E \otimes F$ may be written in a unique way as

$$\mathbf{x} = \sum_j \xi_j \otimes \mathbf{f}_j, \quad \xi_j \in E$$

If G is any vector space over the field of scalars C then $G \approx G \otimes C$. The isomorphism is $g \to g \otimes 1$ since $\{1\}$ is a basis of the vector space C over C.

Suppose now that E and F are topological vector spaces. We want to define a topology in $E \otimes F$. In general, there are several distinct such topologies. However, if F is finite-dimensional, then there is a unique topology defined on $E \otimes F$, the topology of coordinate-wise convergence. Let $\{f_j\}$, $j = 1, \ldots, n$ be a basis of F. Every element $\mathbf{x} \in E \otimes F$ can be written as

$$\mathbf{x} = \sum_{j=1}^{n} \xi_j \otimes \mathbf{f}_j, \quad \xi_j \in E.$$

A sequence $\{\mathbf{x}_k\}$ in $E \otimes F$ converges to zero, $\mathbf{x}_k \underset{k}{\to} 0$, if $\xi_j^k \underset{k}{\to} 0$ for all j. This topology is independent of the basis of F. Every "good" property of E is also possessed by $E \otimes F$. If E is locally convex, reflexive, or complete, then $E \otimes F$ is also locally convex, reflexive, or complete, respectively.

Ket G be a given vector space. If $\beta: E \times F \to G$ is a given bi-linear map, then there is a unique linear map $u: E \otimes F \to G$ such that

$$\beta(x, y) = u(\mathbf{x} \otimes \mathbf{y})$$

for every $(\mathbf{x}, \mathbf{y}) \in E \times F$. Conversely, if $u: E \otimes F \to G$ is a given linear map, then there is a unique bi-linear map $\beta: E \times F \to G$ such that the above relation holds:

$$\begin{array}{ccc} & E \times F & \\ \swarrow & & \searrow \beta \\ E \otimes F & \underset{u}{\to} & G \end{array}$$

If u is given, then β is immediately defined by the above formula. If β is given, the above formula defines u on elements of the form $\mathbf{x} \otimes \mathbf{y}$, and, since every member of $E \otimes F$ is a finite linear combination of such elements, u is defined on $E \otimes F$. This important result demonstrates that the main use of the tensor product is the linearization of bi-linear maps: every bi-linear map $\beta: E \times F \to G$ may be replaced by a linear map $u: E \otimes F \to G$. If E and F are topological vector spaces and F is finite-dimensional, then u is continuous if and only if β is continuous.

Suppose that E and F are topological vector spaces and F is finite-dimensional. Then

$$(E \otimes F)' \approx E' \otimes F',$$

i.e., the dual of $E \otimes F$ is canonically isomorphic and topologically equivalent with the tensor product of the duals of E and F. The topologies here are the strong dual topologies. An element $\mathbf{z} \in E \otimes F$ may be written as

$$\mathbf{z} = \sum_i c_i \mathbf{e}_i \otimes \mathbf{f}_i$$

and an element $\mathbf{z}' \in (E \otimes F)'$ may be written as

$$\mathbf{z}' = \sum_j d'_j \mathbf{e}'_j \otimes \mathbf{f}'_j.$$

The duality product $\langle \mathbf{z}, \mathbf{z}' \rangle$ is given by

$$\langle \mathbf{z}, \mathbf{z}' \rangle = \sum_{i,j} c_i d'_j \langle \mathbf{e}_i, \mathbf{e}'_j \rangle \langle \mathbf{f}_i, \mathbf{f}'_j \rangle.$$

If $\{\mathbf{e}_\alpha\}$, $\{\mathbf{f}_\beta\}$ are bases of E and F respectively, and $\{\mathbf{e}'_\alpha\}$, $\{\mathbf{f}'_\beta\}$ are the corresponding dual bases, then

$$\mathbf{z} = \sum c_{\alpha\beta} \mathbf{e}_\alpha \otimes \mathbf{f}_\beta$$

$$\mathbf{z}' = \sum c'_{\alpha\beta} \mathbf{e}'_\alpha \otimes \mathbf{f}'_\beta$$

and

$$\langle \mathbf{z}, \mathbf{z}' \rangle = \sum c_{\alpha\beta} c'_{\alpha\beta}.$$

Here we used the Kronecker relation $\langle \mathbf{e}_i, \tilde{\mathbf{e}}'_j \rangle = \delta_{ij}$.

Suppose that E and F are topological vector spaces and F is finite-dimensional. Then

$$\mathscr{L}_b(E; F) \approx E' \otimes F$$

where $\mathscr{L}_b(E; F)$ is the space of all continuous linear maps from E to F with the topology of bounded convergence. In particular, $\mathscr{L}(E, C) \approx E' \otimes C \approx E'$. Every element $\mathbf{e}' \otimes \mathbf{f}$ defines a continuous linear map from E into F: $\mathbf{e} \to \langle \mathbf{e}', \mathbf{e} \rangle\ \mathbf{f}$; the element $\sum_\alpha \mathbf{e}'_\alpha \otimes \mathbf{f}_\alpha$ defines the map

$\mathbf{e} \to \sum_\alpha \langle \mathbf{e}'_\alpha, \mathbf{e} \rangle \mathbf{f}_\alpha$; this gives the correspondence between $E' \otimes F$ and $\mathscr{L}(E; F)$.

Vector-valued Particles

Let $\mathbf{F}$ be a finite-dimensional vector space. In this section we shall try to find all the $\mathbf{F}$-valued particles, i.e., all the Hilbert spaces $\mathscr{H}$ with continuous injection in the space $\mathscr{D}'(V; \mathbf{F})$ of $\mathbf{F}$-valued distributions in the universe V. We shall follow a similar procedure as in the case of scalar particles, using the results of the previous section on tensor products.

The locally convex topological vector space E is, in this case,

$$E = \mathscr{D}'(V; \mathbf{F}) = \mathscr{L}_b(\mathscr{D}(V); \mathbf{F}) \approx \mathscr{D}'(V) \otimes \mathbf{F}.$$

Every element of E is of the form $\sum c_i T_i \mathbf{f}_i$, where $T_i \in \mathscr{D}'(V)$ are scalar-valued distributions and $\mathbf{f}_i \in \mathbf{F}$. Its value on an element $\varphi \in \mathscr{D}(V)$ is $\sum c_i \langle T_i, \varphi \rangle \mathbf{f}_i \in \mathbf{F}$.

According to the properties of tensor products, the dual E' of E is

$$E' = \mathscr{D}(V) \otimes \mathbf{F}' \approx \mathscr{D}(V; \mathbf{F}'),$$

i.e., E' is the space of infinitely differentiable functions with compact support in V and with values in the dual $\mathbf{F}'$ of $\mathbf{F}$. Let $\{\mathbf{f}_i\}$ be a basis of $\mathbf{F}$ and $\{\mathbf{f}'_i\}$ the corresponding dual basis of $\mathbf{F}'$. An element $\boldsymbol{\varphi}$ of E' is of the form

$$\boldsymbol{\varphi} = \sum \varphi_i \mathbf{f}'_i,$$

where $\varphi_i \in \mathscr{D}(V)$ are scalar-valued testing functions, and an element $\mathbf{T}$ of E is of the form

$$\mathbf{T} = \sum T_j \mathbf{f}_j$$

where $T_j \in \mathscr{D}'(V)$ are scalar-valued distributions. The value of the distribution $\mathbf{T}$ on the testing function $\boldsymbol{\varphi}$ is

$$\langle \mathbf{T}, \boldsymbol{\varphi} \rangle = \sum \langle T_i, \varphi_i \rangle.$$

Here we used the Kronecker relation $\langle \mathbf{f}_i, \mathbf{f}'_j \rangle = \delta_{ij}$.

In order to find the space $\mathfrak{H}$ of Hilbert spaces $\mathscr{H}$ with continuous injection in E', we must look for the canonically isomorphic space $\overline{\mathscr{L}}_+(E'; E)$ of positive anti-kernels from E' into E. The first step is to find the space $\mathscr{L}_b(E'; E)$ of continuous linear maps from E' into E.

Using the theorem of kernels and the properties of tensor products we obtain

$$\mathscr{L}_b(E'; E) = \mathscr{L}_b(\mathscr{D} \otimes \mathbf{F}'; \mathscr{D}' \otimes \mathbf{F})$$
$$\approx \mathscr{L}_b(\mathscr{D}, \mathscr{D}') \otimes \mathbf{F} \otimes \mathbf{F} \approx \mathscr{D}'(V \times V) \otimes \mathbf{F} \otimes \mathbf{F}.$$

In terms of a basis $\{\mathbf{f}_i\}$ of $\mathbf{F}$, an element $\mathbf{K}$ of $\mathscr{L}_b(E'; E)$ is given by

$$\mathbf{K} = \sum K_{ij}\mathbf{f}_i \otimes \mathbf{f}_j$$

where (K_{ij}) is a square matrix of scalar distributions of two variables. The value of $\mathbf{K}$ on an element $\boldsymbol{\varphi} = \sum \varphi_i \mathbf{f}'_i$ of $E' = \mathscr{D} \otimes \mathbf{F}'$ is the element $\mathbf{K} \cdot \boldsymbol{\varphi}$ of $E = \mathscr{D}' \otimes \mathbf{F}$ given by

$$\mathbf{K} \cdot \boldsymbol{\varphi} = \sum T_i \mathbf{f}_i \in \mathscr{D}' \otimes \mathbf{F}$$

where

$$T_i = \sum_j K_{ij} \cdot \varphi_j \in \mathscr{D}'.$$

Thus the elements $\mathbf{K}$ of $\mathscr{L}_b(E'; E)$ are in one-to-one correspondence with square matrices of scalar distributions of two variables.

Now from the linear map $\varphi \to K \cdot \varphi$ we must obtain an anti-linear map. In the case of scalar particles, this was done by taking instead $\varphi \to K \cdot \overline{\varphi}$, since $\varphi \in \mathscr{D}(V)$ is scalar-valued and the complex conjugate $\overline{\varphi}$ has a meaning. However, in the case of vector-valued particles, an element $\boldsymbol{\varphi} \in \mathscr{D}(V; \mathbf{F}')$ is of the form $\boldsymbol{\varphi} = \sum \varphi_i \mathbf{f}'_i$, where $\mathbf{f}'_i \in \mathbf{F}'$ and complex conjugation here has no meaning. In order to overcome this difficulty we shall introduce the concept of an anti-space.

Definition. Let F be a topological vector space over the field C of complex numbers. An *anti-space* $\overline{F}$ of F is the pair of a topological vector space $\overline{F}$ and anti-isomorphism "bar" from F onto $\overline{F}$, i.e., a one-to-one bicontinuous map $f \to \overline{f}$ such that $\overline{f+g} = \overline{f} + \overline{g}$, $\overline{\lambda f} = \overline{\lambda}\overline{f}$.

The anti-space $\overline{F}$ of F is unique up to an isomorphism. There is no special advantage in using only a particular realization of $\overline{F}$. We present here some examples of such realizations.

(1) Let F be a given topological vector space. The anti-space $\overline{F}$ is identical with F as a set, it has the same topology, the same law of addition, but the law of scalar multiplication is different:

$$(\lambda f)_{\overline{\mathbf{F}}} = (\overline{\lambda} f)_{\mathbf{F}}, \text{ for every } \lambda \in C \text{ and } f \in F.$$

The bar operation is the identity.

(2) Let G' be the dual of a given topological vector space G. The anti-space $\overline{G'}$ of G' is the anti-dual of G, i.e., the space of continuous anti-linear functionals on G. If $g' \in G'$ with $g'\colon g \to \langle g', g\rangle$, then $\overline{g'}$: $g \to \langle \overline{g', g}\rangle$.

(3) In the spaces L^2, $\mathscr{D}$, or $\mathscr{D}'$ there is an internal conjugation. The corresponding anti-spaces are the same as the original and the bar operation is the usual conjugation:

$$f \to \bar{f}, \quad f \in L^2$$

$$\varphi \to \bar{\varphi}, \quad \varphi \in \mathscr{D}$$

$$T \to \bar{T}, \quad T \in \mathscr{D}'.$$

(4) Let $\mathscr{H}$ be a Hilbert space. The dual $\mathscr{H}'$ is a realization of the anti-space $\overline{\mathscr{H}}$. If $h \in \mathscr{H}$ then $\bar{h} \in \mathscr{H}' = \overline{\mathscr{H}}$ with $\langle \bar{h}, k\rangle = (k|h)_{\mathscr{H}}$ for every $k \in \mathscr{H}$. Notice that

$$\langle \overline{\lambda h}, k\rangle = (k\,|\,\lambda h)_{\mathscr{H}} = \bar{\lambda}(k|h)_{\mathscr{H}} = \bar{\lambda}\,\langle \bar{h}, k\rangle$$

so that $\overline{\lambda h} = \bar{\lambda}\,\bar{h}$.

Some of the properties of anti-spaces are:

(a) $\overline{E \otimes F} \approx \bar{E} \otimes \bar{F}$, with the correspondence

$$\sum \overline{c_i e_i \otimes f_i} \in \overline{E \otimes F} \leftrightarrow \sum \bar{c}_i \bar{e}_i \otimes \bar{f}_i \in \bar{E} \otimes \bar{F}$$

(b) $(\bar{F})' \approx (F')^-$.

(c) $\bar{\bar{F}} \approx F$.

Let us return now to the problem of finding the continuous anti-linear maps from $E' = \mathscr{D}(V;\, F')$ into $E = \mathscr{D}'(V, F)$. With the introduction on anti-spaces this problem becomes simple if we notice that a map from E' into E is anti-linear if and only if the corresponding map from $\overline{E'}$ into E is linear. Thus $\overline{\mathscr{L}}_b(E';\, E) \approx \mathscr{L}_b(\overline{E'};\, E)$. Using the properties of anti-spaces

$$\overline{E'} \approx \overline{\mathscr{D}(V) \otimes F'} \approx \overline{\mathscr{D}(V)} \otimes \overline{F'} \approx \mathscr{D}(V) \otimes \overline{F'} \approx \mathscr{D}(V;\overline{F'}).$$

In terms of a basis $\{f_i\}$ in F and the corresponding bases in F' and $\overline{F'}$, if

$$\varphi = \sum \varphi_i f' \in E' = \mathscr{D}(V;F'), \quad \varphi_i \in \mathscr{D}(V),$$

then

$$\bar{\varphi} = \sum \overline{\varphi_i f'} \in \overline{E'} = \mathscr{D}(V;\overline{F'}), \quad \bar{\varphi}_i \in \overline{\mathscr{D}(V)} = \mathscr{D}(V).$$

Using again the theorem of kernels and the properties of tensor products we obtain

$$\mathscr{L}_b(\overline{E'}; E) \approx \mathscr{L}_b(\mathscr{D} \otimes \overline{F'}; \mathscr{D}' \otimes F)$$

$$\approx \mathscr{L}_b(\mathscr{D}; \mathscr{D}') \otimes F \otimes \overline{F} \approx \mathscr{D}'(V \times V) \otimes F \otimes \overline{F},$$

and finally the space $\overline{\mathscr{L}}_b(E'; E)$ of continuous anti-linear maps from E' into E is given by

$$\overline{\mathscr{L}}_b(E'; E) \approx \mathscr{D}'(V \times V) \otimes F \otimes \overline{F}.$$

The last step in our search is to find the subspace $\overline{\mathscr{L}}_+(E', E)$ of $\overline{\mathscr{L}}_b(E'; E)$ consisting of positive elements.

Definition. An anti-kernel $\varphi \to K \cdot \overline{\varphi}$ in $\overline{\mathscr{L}}_b(E'; E)$ defined by a distribution $\mathbf{K}_{x\xi} \in \mathscr{D}'(V \times V) \otimes \mathbf{F} \otimes \overline{\mathbf{F}}$ is *super-positive* if

$$\langle \mathbf{K}, \boldsymbol{\varphi} \otimes \overline{\boldsymbol{\varphi}} \rangle \geqslant 0$$

for every $\boldsymbol{\varphi} \in \mathscr{D}(V; \mathbf{F}')$. In terms of a basis $\{\mathbf{f}_i\}$ in $\mathbf{F}$, $\boldsymbol{\varphi} = \sum \varphi_i \mathbf{f}'_i$, $\mathbf{K} = \sum K_{ij} \mathbf{f}_i \otimes \mathbf{f}_i$ with $K_{ij} \in \mathscr{D}'(V \times V)$

and
$$\langle \mathbf{K}, \boldsymbol{\varphi} \otimes \overline{\boldsymbol{\varphi}} \rangle = \sum_{ij} \langle K_{ij}, \varphi_i \otimes \overline{\varphi}_j \rangle \geqslant 0.$$

This is a combination of the notion of positivity for matrices and kernels.

We may state now the final results of this section in the following theorem:

Theorem. The space $\mathfrak{H}$ of Hilbert spaces $\mathscr{H}$ with continuous injection in $\mathscr{D}'(V; F)$ is canonically isomorphic with the subspace of $\mathscr{D}'(V \times V) \otimes F \otimes \overline{F}$ consisting of distributions K on $V \times V$ with values in $F \otimes \overline{F}$ such that

$$\langle K, \varphi \otimes \overline{\varphi} \rangle \geqslant 0$$

for every $\varphi \in \mathscr{D}(V; F')$. The relation between such a K and the corresponding $\mathscr{H}$ is given by the formula

$$\langle T, \varphi \rangle = (T | K \cdot \overline{\varphi})_{\mathscr{H}}$$

for every $T \in \mathscr{H} \subset \mathscr{D}'(V; F)$, and every $\varphi \in \mathscr{D}(V; F')$. The element $K \cdot \overline{\varphi}$ of $\mathscr{H}$ is defined by

$$\langle K \cdot \overline{\varphi}, \psi \rangle = \langle K, \psi \otimes \overline{\varphi} \rangle, \quad \psi \in \mathscr{D}(V; F').$$

If we let $T = K \cdot \overline{\psi}$ we obtain

$$\langle K, \varphi \otimes \overline{\psi} \rangle = (K \cdot \overline{\psi} | K \cdot \overline{\varphi})_{\mathscr{H}}, \quad \varphi, \psi \in \mathscr{D}(V; F').$$

These are exactly the same formulas as in the case of scalar particles. For convenience, we present here a list of formulas in terms of a basis $\{f_i\} \in F$:

$$\varphi = \sum \varphi_i f'_i \in E' = \mathscr{D}(V : F'), \quad \varphi_i \in \mathscr{D}(V),$$
$$T = \sum T_i f_i \in E = \mathscr{D}'(V ; F), \quad T_i \in \mathscr{D}'(V),$$
$$\langle T, \varphi \rangle = \sum \langle T_i, \varphi_i \rangle,$$
$$K = \sum_{i,j} K_{ij} f_i \otimes \bar{f}_j \in \mathscr{D}'(V \times V) \otimes F \otimes \bar{F}, \quad K_{ij} \in \mathscr{D}'(V \times V),$$
$$K \cdot \bar{\varphi} = \sum S_i f_i \in \mathscr{D}'(V ; F),$$
$$S_i = \sum K_{ij} \cdot \bar{\varphi}_j \in \mathscr{D}'(V),$$
$$\langle K, \varphi \otimes \bar{\psi} \rangle = \sum \langle K_{ij}, \varphi_i \otimes \bar{\psi}_j \rangle.$$

Order Relations in Vector Spaces and Positivity of Anti-kernels

Let E be a vector space over the reals R. We want to define an order relation $\geqslant$ in E which is compatible with the vector structure of E. This means that the order relation should be invariant under translation and positive homotheties, so that

$$x \geqslant y \Leftrightarrow \begin{cases} x - a \geqslant y - a & \text{for all } a \in E, \text{ and} \\ \lambda x \geqslant \lambda y & \text{for all } \lambda > 0. \end{cases}$$

In order to define an order relation in E it is sufficient to know the set of "positive" elements $\{x : x \geqslant 0\}$, because then we may set

$$x \geqslant y \Leftrightarrow x - y \geqslant 0.$$

Recalling that $\Gamma \subset E$ is a convex cone if

$$\begin{aligned} &x, y \in \Gamma \Rightarrow x + y \in \Gamma, \text{ and} \\ &x \in \Gamma \Rightarrow \lambda x \in \Gamma \qquad \text{for all } \lambda > 0, \end{aligned}$$

we may easily verify the following theorem:

Theorem. The set of convex cones $\Gamma \subset E$ such that $\Gamma \cap (-\Gamma) = \{0\}$ is in one-to-one correspondence with the set of order relations compatible with the vector structure in E. This correspondence is given by

$$x \geqslant y \Leftrightarrow x - y \in \Gamma.$$

The order relation is closed if and only if the cone Γ is closed.

The condition $\Gamma \cap (-\Gamma) = \{0\}$ means that $x \in \Gamma$, $-x \in \Gamma$ implies that $x = 0$.

An order relation in an affine space is defined naturally by the order relation in the associated vector space: $x \geqslant y \Leftrightarrow \overrightarrow{x - y} \in \Gamma$.

The closed convex cone Γ defines, under certain conditions, a dual order relation in the dual E' of E: an element $e' \in E'$ is *positive* if $\langle e', e \rangle \geqslant 0$ for all $e \in \Gamma$.

Let
$$\Gamma' = \{e' : \langle e', e \rangle \geqslant 0 \text{ for all } e \in \Gamma\}.$$

Clearly, Γ' is a convex cone which is weakly closed, therefore Γ' is strongly closed so that Γ' is a closed convex cone. Therefore Γ' defines an order relation in E' provided the condition $\Gamma' \cap (-\Gamma') = \{0\}$ is satisfied. In order to find the corresponding condition in terms of Γ we introduce the concept of polar sets.

Definition. Let $A \subset E$. The *polar set* $A^0 \subset E'$ of A is

$$A^0 = \{e' : e' \in E', \langle e', e \rangle \geqslant -1 \text{ for all } e \in A\}.$$

Note that the polar set of any set is always convex and weakly closed. Furthermore $(A_1 \cup A_2)^0 = A_1^0 \cap A_2^0$. If A is a convex cone, then the condition $\langle e', e \rangle \geqslant -1$ for all $e \in A$ implies that $\langle e', e \rangle \geqslant 0$ for all $e \in A$. Thus for our convex cone Γ we have $\Gamma' = \Gamma^0$.

Theorem. $\Gamma^0 \cap (-\Gamma^0) = \{0\} \Leftrightarrow$ the space spanned by Γ is dense in E.

We shall prove only that if the space spanned by Γ is dense in E then $\Gamma^0 \cap (-\Gamma^0) = \{0\}$. Suppose that $e' \in \Gamma^0 \cap (-\Gamma^0)$. Then e' is both positive and negative on Γ and hence it is zero on Γ. Since e' is a continuous linear functional on E and the space spanned by Γ is dense in E, e' is zero on E. Hence $e' = 0$.

Corollary. If the space spanned by Γ is dense in E then Γ^0 defines a dual order relation in E'.

Since Γ is a closed convex cone, $\Gamma^{00} = \Gamma$. Hence the order relation in E may be defined by the order relation in E': $e \in E$ is positive $\Leftrightarrow \langle \tilde{\ }, e \rangle \geqslant 0$ for all $e' \in \Gamma^0$.

In what follows we shall assume that Γ is a closed convex cone in E such that $\Gamma \cap (-\Gamma) = \{0\}$ and the space spanned by Γ is dense in E. Thus Γ defines an order relation in E and Γ^0 defines the dual order relation in E'.

Let us turn now to complex vector spaces. Let E be a vector space over the field of C of complex numbers. We shall assume that E is the complexification of a vector space E_0 over the reals, i.e.,

$$E = E_0 + iE_0 = \{x + iy : x, y \in E_0\};$$

thus E is the direct sum of E_0 and iE_0. Given the space E and the real subspace $E_0 \subset E$, there is a bar operation $z \to \bar{z}$ (complex conjugation) defined in E with the property $\bar{\bar{z}} = z$. This operation is a canonical anti-isomorphism, i.e., an anti-linear bi-continuous one-to-one mapping from E onto E.

The subspace E_0 is the subspace of elements of E which are self-conjugate, i.e., $E_0 = \{z : z \in E, \bar{z} = z\}$. Every element $z \in E$ can be written in a unique way as

$$z = \frac{z + \bar{z}}{2} + i\frac{z - \bar{z}}{2i}$$

where $z + \bar{z}/2$ and $z - \bar{z}/2i$ belong to E_0. Thus the bar structure in E defines (and is defined by) the real subspace E_0.

The dual E' of E has a bar structure also. In fact the relation

$$\overline{\langle z', z \rangle} = \langle \bar{z}', \bar{z} \rangle$$

or

$$\overline{\langle z', \bar{u} \rangle} = \langle \bar{z}', u \rangle, \qquad z' \in E, u \in E,$$

defines $\bar{z}'$. An element z' of E' is *real* if it is self-conjugate, i.e., if it takes real values on E_0. The real subspace of E' (consisting of real elements) is canonically isomorphic to the (real) dual E_0' of E_0. Thus

$$E' = E_0' + iE_0'.$$

Let Γ be the cone of positive elements of E_0, i.e., Γ is a closed convex cone in E_0 such that $\Gamma \cap (-\Gamma) = \{0\}$ and the (real) space spanned by Γ is dense in E_0. Then $\Gamma' \subset E_0'$ is defined by

$$\Gamma' = \{z' : z' \in E_0', \langle z', z \rangle \geqslant 0 \text{ for all } z \in \Gamma\},$$

and an element of E' is *positive* if it takes positive values on the positive elements of E_0.

Let us apply these results to the spaces we are concerned with. Consider first the spaces $\mathscr{D}$ and $\mathscr{D}'$. The conjugate $\bar{T}$ of T is defined by

$$\overline{\langle T, \varphi \rangle} = \langle \bar{T}, \bar{\varphi} \rangle.$$

Definition. An element $T \in \mathscr{D}'$ is *real* if $\langle T, \varphi \rangle$ is real for every real $\varphi \in \mathscr{D}$. T is *positive* if $\langle T, \varphi \rangle \geqslant 0$ for every positive $\varphi \in \mathscr{D}$.

Now let us consider the space $F \otimes \bar{F}$, where F is a finite-dimensional vector space. The bar structure in $F \otimes \bar{F}$ is defined as follows. The image of an element $\sum c_\nu f_\nu \otimes \bar{g}_\nu \in F \otimes \bar{F}$ under the bar operation is $\sum \bar{c}_\nu \bar{f}_\nu \otimes g_\nu$ which we identify with $\sum \bar{c}_\nu g_\nu \otimes \bar{f}_\nu$.
Thus

$$\overline{F \otimes \bar{F}} = \bar{F} \otimes F = F \otimes \bar{F}.$$

In terms of a basis $\{f_i\}$ in F the bar conjugate of an element $\sum c_{ij} f_i \otimes \bar{f}_j \in F \otimes \bar{F}$ is $\sum \overline{c_{ij}} \bar{f}_i \otimes f_j$, and this element is invariant under conjugation, i.e.,

$$\sum c_{ij} f_i \otimes \bar{f}_j = \sum \overline{c_{ij}} \bar{f}_i \otimes f_j = \sum \overline{c_{ji}} f_i \otimes \bar{f}_j$$

if and only if $c_{ij} = \overline{c_{ji}}$. Thus the self-conjugate elements are represented by hermitian matrices.

Definition. An element of $F \otimes \bar{F}$ is *real* if, for any basis of F, it is represented by a hermitian matrix.

An equivalent definition is the following: an element of $F \otimes \bar{F}$ is real if it is a linear combination of squares with real coefficients, i.e., if it is of the form $\sum \lambda_\nu g_\nu \otimes \bar{g}$ with λ_ν real. Let

$$\sum c_{ij} f_i \otimes \bar{f}_j = \sum \lambda_\nu g_\nu \otimes \bar{g}_\nu$$

and

$$g_\nu = \sum d_{\nu i} \ f_i.$$

Then

$$c_{ij} = \sum \lambda_\nu d_{\nu i} \overline{d_{\nu j}}$$

where $c_{ij} = \bar{c}_{ji}$ if λ_v is real. Conversely, if $c_{ij} = \bar{c}_{ji}$, the theorem of decomposition of a hermitian form proves that there is an expression for c_{ij} like the one above, with λ_v real.

Definition. An element of $F \otimes \bar{F}$ is *positive* if, for any basis of F, it is represented by a positive hermitian matrix. Or an element of $F \otimes \bar{F}$ is positive if it is a linear combination of squares with positive coefficients.

Still another equivalent definition of positivity is possible. An element $A = \sum c_\nu f_\nu \otimes \bar{g}_\nu \in F \otimes \bar{F}$ defines a linear–anti-linear form on $F' \times F'$

as follows:

$$A(u', v') = \sum c_\nu \langle u', f_\nu \rangle \langle \overline{v', g_\nu} \rangle.$$

This form is positive hermitian $\Leftrightarrow$ the element $\sum c_\nu f_\nu \otimes \overline{g_\nu}$ is positive.

The above definition of positive elements provides an order relation in $F \otimes \overline{F}$ since there is an order relation for hermitian matrices. Finally we have the following definition of positive elements in the dual $F' \otimes \overline{F'}$ of $F \otimes \overline{F}$.

Definition. An element of $F' \otimes \overline{F'}$ is positive if it takes positive values on every positive element of $F \otimes \overline{F}$.

We shall discuss now the notion of positivity for anti-kernels defined by elements $K_{x, \xi} \in \mathscr{D}'(V \times V) \otimes F \otimes \overline{F}$. Here we have two definitions of positivity, both of which seem to be natural.

Definition. An anti-kernel defined by an element $K \in \mathscr{D}'(V \times V) \otimes F \otimes \overline{F}$ is called

(a) *positive* if for every $\varphi \in \mathscr{D}(V)$
$\langle K, \varphi \otimes \overline{\varphi} \rangle \geqslant 0$ in the natural order of $F \otimes \overline{F}$,
(here $\langle K, \varphi \otimes \overline{\varphi} \rangle \in F \otimes \overline{F}$).

(b) *super-positive* if for every $\psi \in \mathscr{D}(V; F')$
$\langle K, \psi \otimes \overline{\psi} \rangle \geqslant 0$ in R
(here $\langle K, \psi \otimes \overline{\psi} \rangle \in R$).

In the previous section we required that the anti-kernel K corresponding to the Hilbert space $\mathscr{H} \subset \mathscr{D}'(V; F)$ be superpositive (there, however, we simply called K positive).

In terms of a basis $\{f_i\}$ in F an element $K \in \mathscr{D}'(V \times V) \otimes F \otimes \overline{F}$ is given by $K = \sum K_{ij} f_i \otimes \overline{f_j}$, where $K_{ij} \in \mathscr{D}'(V \times V)$ and an element $\psi \in \mathscr{D}(V : F')$ is given by $\psi = \sum \varphi_i f_i'$ where $\varphi_i \in \mathscr{D}(V)$. Now the above two definitions become

(a) K is positive if for every $\varphi \in \mathscr{D}(V)$ the matrix
$((\langle K_{ij}, \varphi \otimes \overline{\varphi} \rangle))$ is positive hermitian.

(b) K is super-positive if for every $\{\varphi_i\} \subset \mathscr{D}(V)$
$\sum \langle K_{ij}, \varphi_i \otimes \overline{\varphi_j} \rangle \geqslant 0$ in R.

Taking $\varphi_i = c_i \varphi_i$, $c_i \in C$, it is easy to see that super-positivity implies positivity. It is believed, however, that the two notions are not equivalent and a counter-example is welcomed.

Up to now, the notions of order relation in a vector space and of (simple) positivity of anti-kernels were not used. They will be needed, however, later on. It will be shown that for certain anti-kernels the two notions of positivity are equivalent.

CHAPTER 3

Universal Particles and the Translation Invariance: Kernel Simplification

Tensor Products of Distributions

Let X and Y be C^∞-manifolds, and let $\mathscr{D}_x$ and $\mathscr{D}_y$ be the spaces of testing functions on X and Y, respectively. For any $u(x) \in \mathscr{D}_x$ and $v(y) \in \mathscr{D}_y$, $u(x)v(y)$ will be infinitely differentiable in x and y, and the support of $u(x)v(y)$ is compact, since the cartesian product of compact sets is compact. Thus, $u(x)v(y)$ is an element of $\mathscr{D}_{x,y}$, the space of testing functions on $X \times Y$. It follows that

$$\mathscr{D}_x \otimes \mathscr{D}_y \subset \mathscr{D}_{x,y}.$$

Furthermore, it can be shown that $\mathscr{D}_x \otimes \mathscr{D}_y$ is a proper dense subset of $\mathscr{D}_{x,y}$.

The same result holds for the spaces of distributions: $\mathscr{D}'_x \otimes \mathscr{D}'_y$ is a proper dense subset of $\mathscr{D}'_{x,y}$. We shall not give the proof of this here. We shall only prove that the tensor product of two distributions in $\mathscr{D}'_x$ and $\mathscr{D}'_y$ defines an element in $\mathscr{D}'_{x,y}$.

Theorem. Given $S_x \in \mathscr{D}'_x$ and $T_y \in \mathscr{D}'_y$, there exists one and only one distribution $S_x \otimes T_y \in \mathscr{D}'_{x,y}$ such that

$$\langle S_x \otimes T_y, u(x) \otimes v(y)\rangle = \langle S_x, u(x)\rangle \langle T_y, v(y)\rangle$$

for all $u(x) \in \mathscr{D}_x$, $v(y) \in \mathscr{D}_y$.

Proof. Iniqueness: Suppose that there are two different distributions $U_1, U_2 \in \mathscr{D}_{x,y}$ such that

$$\langle U_1, u(x) \otimes v(y)\rangle = \langle S_x, u(x)\rangle \langle T_y, v(y)\rangle$$

$$\langle U_2, u(x) \otimes v(y) = \langle S_x, u(x)\rangle \langle T_y, v(y)\rangle.$$

Obviously, then, U_1 and U_2 coincide on $\mathscr{D}_x \otimes \mathscr{D}_y$. But since $\mathscr{D}_x \otimes \mathscr{D}_y$ is dense in $\mathscr{D}_{x,y}$ and because of the continuity of the duality products, it follows that the values of U_1 and U_2 coincide on all of $\mathscr{D}_{x,y}$. Thus

$U_1 = U_2$. Existence: For any $\varphi(x, y) \in \mathscr{D}_{x,y}$, we want to see if $\langle S_x \otimes T_y, \varphi(x, y)\rangle$ has a meaning. For each fixed value of x, $\varphi(x, y) \in \mathscr{D}_y$, and $\langle T_y, \varphi(x, y)\rangle$ is a function of x which obviously has a compact support. It can be shown that this function of x is infinitely differentiable in x. Thus we have

$$\langle T_y, \varphi(x, y)\rangle \in \mathscr{D}_x.$$

Now we may compute

$$\langle S_x, \langle T_y, \varphi(x, y)\rangle\rangle$$

and thus for every $\varphi \in \mathscr{D}_{x,y}$ we have obtained a number. Define

$$\langle S_x \otimes T_y, \varphi(x, y)\rangle = \langle S_x, \langle T_y, \varphi(x, y)\rangle\rangle.$$

This gives us a continuous linear functional on $\mathscr{D}_{x,y}$, but the continuity will not be proved here. Now if $\varphi(x, y) = u(x)v(y)$ for $u(x) \in \mathscr{D}_x$, $v(y) \in \mathscr{D}_y$, we have

$$\langle T_y, u(x) \otimes v(y)\rangle = u(x)\langle T_y, v(y)\rangle \in \mathscr{D}_x.$$

and the application of the operator S_x gives

$$\langle S_x \otimes T_y, u(x) \otimes v(y)\rangle = \langle S_x, \langle T_y, u(x) \otimes v(y)\rangle\rangle$$
$$= \langle S_x, u(x)\rangle \langle T_y, v(y)\rangle.$$

This completes the proof of the existence. Another way is to define

$$\langle S_x \otimes T_y, \varphi(x, y)\rangle = \langle T_y, \langle S_x, \varphi(x, y)\rangle\rangle$$

but because of the uniqueness the two definitions are equivalent, i.e.,

$$\langle T_y, \langle S_x, \varphi(x, y)\rangle\rangle = \langle S_x, \langle T_y, \varphi(x, y)\rangle\rangle.$$

The same theorem is true for vector valued distributions when the range space of one vector distribution has a finite dimension:

Theorem. Given $\mathbf{S}_x \in \mathscr{D}'(X; \mathbf{E})$, $\mathbf{T}_y \in \mathscr{D}'(Y; \mathbf{F})$, and either $\mathbf{E}$ or $\mathbf{F}$ has a finite dimension, then there exists one and only one distribution $\mathbf{S}_x \otimes \mathbf{T}_y \in \mathscr{D}'(X \times Y; \mathbf{E} \times \mathbf{F})$ such that

$$\langle \mathbf{S}_x \otimes \mathbf{T}_y, u(x) \otimes v(y)\rangle = \langle \mathbf{S}_x, u(x)\rangle \otimes \langle \mathbf{T}_y, v(y)\rangle.$$

Proof. The uniqueness proof is the same as before.
Existence: If both $\mathbf{E}$ and $\mathbf{F}$ are finite dimensional it is simple. For $\mathbf{S}_x \in \mathscr{D}'_x \otimes \mathbf{E}$, $\mathbf{T}_y \in \mathscr{D}'_y \otimes \mathbf{F}$, it is obvious from the previous theorem that $\mathbf{S}_x \otimes \mathbf{T}_y$ can be defined as an element in $\mathscr{D}'_x \otimes \mathscr{D}'_y \otimes \mathbf{E} \otimes \mathbf{F}$.

We know that $\mathscr{D}'_x \otimes \mathscr{D}'_y \subset \mathscr{D}'_{x,y}$, therefore $\mathscr{D}'_x \otimes \mathscr{D}'_y \otimes \mathbf{E} \otimes \mathbf{F} \subset \mathscr{D}'_{x,y} \otimes \mathbf{E} \otimes \mathbf{F}$. Thus $\mathbf{S}_x \otimes \mathbf{T}_y$ is defined in $\mathscr{D}'(X \times Y; \mathbf{E} \otimes \mathbf{F})$ due to the canonical isomorphism of this space with $\mathscr{D}'_{x,y} \otimes \mathbf{E} \otimes \mathbf{F}$.

Suppose that $\mathbf{E}$ is finite-dimensional and $\mathbf{F}$ is not necessarily so. Then $\mathbf{S} \otimes \mathbf{T}_y$ is defined by the equation

$$\langle \mathbf{S}_x \otimes \mathbf{T}_y, \varphi(x, y)\rangle = \langle \mathbf{T}_y, \langle \mathbf{S}_x, \varphi(x, y)\rangle\rangle.$$

For fixed y, $\langle \mathbf{S}_x, \varphi(x, y)\rangle \in \mathbf{E}$, and since $\mathbf{E}$ is finite-dimensional, $\langle \mathbf{S}_x, \varphi(x, y)\rangle \in \mathscr{D}_y \otimes \mathbf{E}$. Now $\mathbf{T}_y \in \mathscr{D}'(Y; \mathbf{F})$, i.e., $\mathbf{T}_y$ defines a map $\mathscr{D}_y \to \mathbf{F}$. Then $\langle \mathbf{T}_y, \langle \mathbf{S}_x, \varphi(x, y)\rangle\rangle$ is a map $\mathscr{D}_y \otimes \mathbf{E} \to \mathbf{F} \otimes \mathbf{E}$. Thus $\mathbf{S}_x \otimes \mathbf{T}_y$ is well defined by the map $\varphi(x, y) \to \langle \mathbf{T}_y, \langle \mathbf{S}_x, \varphi(x, y)\rangle\rangle$, which is linear and continuous.

Remarks. If $\mathbf{E}$ and $\mathbf{F}$ were both ∞-dimensional, then the tensor product $\mathbf{S} \otimes \mathbf{T}$ could not be defined in this way.

The support of a tensor product $\mathbf{S} \otimes \mathbf{T}$ is the cartesian product $A \times B$ of the support A of $\mathbf{S}$ and the support B of $\mathbf{T}$.

The tensor product is continuous, i.e., if $\mathbf{S} \to \mathbf{S}_0$ and $\mathbf{T} \to \mathbf{T}_0$, then $\mathbf{S} \otimes \mathbf{T} \to \mathbf{S}_0 \otimes \mathbf{T}_0$. The proof of this is omitted.

When a basis is chosen, $\mathbf{e}_i \in \mathbf{E}$, $\mathbf{f}_j \in \mathbf{F}$, then the tensor product of $\mathbf{S} = \sum_i S_i \mathbf{e}_i$ and $\mathbf{T} = \sum_j T_j \mathbf{f}_j$ can be written

$$\mathbf{S} \otimes \mathbf{T} = \sum_{i,j} S_i \otimes T_j \, \mathbf{e}_i \otimes \mathbf{f}_j.$$

Convolutions

Let V be a manifold on which there is an internal law of composition $(x, y) \to xy$, i.e., a map from $V \times V$ into V. Assume also that this law of composition is infinitely differentiable. A law of composition called the convolution can be defined on $\mathscr{D}'(V)$, the space of distributions on V.

Definition. Given S and $T \in \mathscr{D}'(V)$, the *convolution* $S * T \in \mathscr{D}'(V)$ is defined by the equation

$$\langle S * T, \varphi\rangle = \langle S_\xi \otimes T_\eta, \varphi(\xi\eta)\rangle$$

for $\varphi \in \mathscr{D}(V)$.

Note. The convolution does not always exist. Conditions under which the convolution does exist will soon be discussed.

The function $\varphi(\xi\eta)$ is an infinitely differentiable function of $(\xi, \eta) \in V \times V$. However, the set $\{(\xi, \eta)\colon \xi\eta \in \text{support of } \varphi \text{ in } V\}$, which is the support of $\varphi(\xi\eta)$ in $V \times V$, is generally not compact (unless $\varphi \equiv 0$), therefore the convolution can not yet be defined.

Example. Let $V = R$, the real line, and let the law of composition be addition. If $\varphi(x_0) \neq 0$, then $\varphi(\xi + \eta) \neq 0$ for $\xi + \eta = x_0$, i.e., the diagonal $\xi + \eta = x_0$ lies in the support of $\varphi(\xi + \eta)$ in $R \times R$. Thus the support of $\varphi(\xi + \eta)$ in $R \times R$ can not be compact (unless $\varphi \equiv 0$).

In order to define a convolution it is necessary to extend the definition of $\langle T, \varphi \rangle$ to certain cases where T is a distribution and φ is an infinitely differentiable function with a support which is not compact.

Theorem. Let T be a distribution with support A and φ be an infinitely differentiable function with support B. If $A \cap B$ is compact, then $\langle T, \varphi \rangle$ can be defined.

Proof. Take a function $\alpha \in \mathscr{D}(V)$ such that $\alpha = 1$ in a neighborhood of $A \cap B$.
$\langle T, \alpha\varphi \rangle$ then has a meaning because $\alpha\varphi$ is a testing function. It must be shown that the result is independent of the choice of α. Suppose that β is another function chosen in the same way as α. Now $\alpha - \beta = 0$ in a neighborhood of $A \cap B$ and the support of $(\alpha - \beta)\,\varphi$ is contained in the complement of $A \cap B$. From the definition of the support of T we have $\langle T, (\alpha - \beta)\,\varphi \rangle = 0$. Therefore we may define

$$\langle T, \varphi \rangle = \langle T, \alpha\varphi \rangle$$

and its value is independent of the choice of α.

Definition. Given a distribution $S \in \mathscr{D}'(V)$ with support $A \subset V$ and a distribution $T \in \mathscr{D}'(V)$ with support $B \subset V$, the supports A and B are said to *permit convolution* if the intersection of $A \times B$ (= support of $S \otimes T$) with the support of $\varphi(\xi\eta)$ in $V \times V$ is compact for every $\varphi \in \mathscr{D}(V)$.

Definition. If E and F are locally compact, a continuous map $E \to F$ is *proper* if any of the following equivalent conditions is satisfied:

(1) The inverse image of every compact set is compact.

(2) The image of every closed set is closed and the inverse image of every point is compact.

(3) If E and F are compactified by adjoining the point ∞ to them (Alexandroff compactification), then the extended map in which ∞ is mapped into ∞ is continuous.

Now we can give another definition for two supports to permit convolution. This definition is equivalent to the one previously stated.

Definition. If the sets A and B are the supports of two distributions, A and B *permit convolution* if any of the following equivalent conditions is satisfied:

(1) For any compact set $K \subset V$, the set $(A \times B) \cap \{(\xi, \eta) : \xi\eta \in K\}$ is compact.

(2) The map $(\xi, \eta) \to \xi\eta$ restricted to $A \times B$ is proper.

Example.

$$\begin{aligned}\langle \delta_{(a)} * \delta_{(b)}, \varphi \rangle &= \langle (\delta_a)_\xi \otimes (\delta_b)_\eta, \varphi(\xi\eta) \rangle \\ &= \langle (\delta_a)_\xi, \langle (\delta_b)_\eta, \varphi(\xi\eta) \rangle \rangle \\ &= \langle (\delta_a)_\xi, \varphi(\xi b) \rangle = \varphi(ab).\end{aligned}$$

Therefore $\delta_{(a)} * \delta_{(b)} = \delta_{(ab)}$.

The following theorem is stated without proof.

Theorem. If we have two converging sequences of distributions

$$S \to S_0$$

$$T \to T_0$$

where

(1) the supports of all the S are contained in the same set A and the supports of all the T are in the same set B.

(2) A and B permit convolution,

then

(i) $S * T \to S_0 * T_0$

(ii) the support of $S_0 * T_0$ is contained in $AB = \{\xi\eta : (\xi, \eta) \in A \times B\}$.

Let V be a Lie group with the product operation $(\xi, \eta) \to \xi\eta$.

Definition. The *internal convolution* of three distributions $R, S, T \in \mathscr{D}'(V)$ is defined by the equation

$$\langle R * S * T, \varphi \rangle = \langle R_\xi \otimes S_\eta \otimes T_\zeta, \varphi(\xi\eta\zeta) \rangle$$

for $\varphi \in \mathscr{D}(V)$.

Definition. The three supports A, B, C of the distributions $R, S, T \in \mathscr{D}'(V)$ permit *internal convolution* if the map $(\xi, \eta, \zeta) \to \xi\eta\zeta$ restricted to $A \times B \times C$ is proper.

Other equivalent definitions are possible.

Theorem. Let $R, S, T, \in \mathscr{D}'(V)$ have the supports A, B and C. If

(1) $(\xi\eta)\zeta = \xi(\eta\zeta)$

(2) A, B, and C permit internal convolution,

then

$$(R * S) * T = R * (S * T) = R * S * T.$$

The proof, which is easy, will not be given.

Theorem. Let e be the unit of the Lie group V. Then

$$\delta_{(e)} * T = T * \delta_{(e)} = T.$$

Proof.

$$\langle \delta_{(e)} * T, \varphi \rangle = \langle (\delta_e)_\xi \otimes T_\eta, \varphi(\xi\eta) \rangle$$

$$= \langle T_\eta, \langle (\delta_e)_\xi, \varphi(\xi\eta) \rangle \rangle$$
$$= \langle T_\eta, \varphi(e\eta) \rangle = \langle T_\eta, \varphi(\eta) \rangle.$$

Theorem. Let $\Lambda_a T$ be the left translation defined by $a \in V$ on the distribution $T \in \mathscr{D}'(V)$ in the Lie group V.
Then

$$\delta_{(a)} * T = \Lambda_a T$$

Similarly for the right translation R_a,

$$T * \delta_{(a)} = R_a T.$$

Proof.

$$\langle \Lambda_a T, \varphi \rangle = \langle T, \Lambda_a^{-1} \varphi \rangle = \langle T_\eta, \varphi(a\eta) \rangle$$
$$= \langle (\delta_{(a)})_\xi \otimes T_\eta, \varphi(\xi\eta) \rangle$$
$$= \langle \delta_{(a)} * T, \varphi \rangle$$

Theorem. Let V be a vector space (addition in V is, of course, commutative). The convolution of $S, T \in \mathscr{D}'(V)$ is commutative.

Proof.

$$\langle S_\xi \otimes T_\eta, \varphi(\xi + \eta) \rangle = \langle S_\eta \otimes T_\xi, \varphi(\eta + \xi)$$
$$= \langle T_\xi \otimes S_\eta, \varphi(\eta + \xi) \rangle = \langle T_\xi \otimes S_\eta, \varphi(\xi + \eta) \rangle.$$

Theorem. Let R be the real line. Then for $T \in \mathscr{D}'(R)$

$$\delta' * T = T * \delta' = T'$$

Proof.

$$\langle \delta' * T, \varphi \rangle = \langle \delta'_\xi \otimes T_\eta, \varphi(\xi + \eta) \rangle$$
$$= \langle T_\eta, \langle \delta'_\xi, \varphi(\xi + \eta) \rangle \rangle$$
$$= \langle T_\eta, -\varphi'(0 + \eta) \rangle = -\langle T_\eta, \varphi'(\eta) \rangle$$

$$= -\langle T, \varphi' \rangle = \langle T', \varphi \rangle$$

Similarly

$$\langle T * \delta', \varphi \rangle = \langle T', \varphi \rangle$$

It also follows that

$$\delta^{(m)} * T = T^{(m)}$$
$$D^p\delta * T = D^pT$$

where D^p is a differential operator of order p with constant coefficients. In more than one dimension we have

$$\Delta\delta * T = \Delta T$$

where Δ may be the Laplacian or the D'Alembertian.

Theorem. For S, T, $\in \mathscr{D}'(R)$ we have

$$(S * T)' = S' * T = S * T'$$

Proof.
$$(S * T)' = \delta' * (S * T) = (\delta' * S) * T = S' * T$$

and similarly

$$(S * T)' = (S * T) * \delta' = S * (T * \delta') = S * T'.$$

It follows that for S, $T \in \mathscr{D}'(V)$, where V is a vector space over the reals with dimension $\geqslant 2$:

$$\frac{\partial^2}{\partial x \partial y}(S * T) = \frac{\partial^2 S}{\partial x \partial y} * T = \frac{\partial S}{\partial x} * \frac{\partial T}{\partial y} = S * \frac{\partial^2 T}{\partial x \partial y} = \text{etc.}$$

Theorem. Let $T \in \mathscr{D}'(V)$ and $\alpha \in \mathscr{D}(V)$ on a vector space V with a Lebesgue measure dx. If the supports of T and α permit convolution, put

$$T * \alpha = \beta.$$

Then β is an infinitely differentiable function given by

$$\beta(x) = \langle T_t, \alpha(x - t) \rangle$$

Proof. For every $\varphi \in \mathscr{D}$

$$\begin{aligned}\langle T * \alpha, \varphi\rangle &= \langle T_\xi \otimes \alpha(\eta), \varphi(\xi + \eta)\rangle = \langle T_\xi, \int \alpha(\eta)\, \varphi(\xi + \eta)\, \mathrm{d}\eta\rangle \\ &= \langle T_\xi, \int \alpha(x - \xi)\, \varphi(x)\, \mathrm{d}x\rangle = \langle T_\xi \otimes \varphi(x), \alpha(x - \xi)\rangle \\ &= \int \varphi(x) \langle T_\xi, \alpha(x - \xi)\rangle\, \mathrm{d}x = \int \varphi(x)\, \beta(x)\, \mathrm{d}x = \langle \beta, \varphi\rangle.\end{aligned}$$

Thus $\beta(x) = \langle T_t, \alpha(x - t)\rangle$. According to the previous theorem

$$\beta^{(m)} = (T * \alpha)^{(m)} = T * \alpha^{(m)}.$$

Since α is infinitely differentiable, $\beta^{(m)} = T * \alpha^{(m)}$ is also a function and it follows that $\beta(x)$ is infinitely differentiable.

Theorem. Let $V = \mathbf{E}$, a vector space with a Lebesgue measure. If f and g are locally integrable functions on $\mathbf{E}$, and the supports of f and g permit convolution, then

$$f * g = h$$

where

$$h(x) = \int f(t)\, g(x - t)\, \mathrm{d}t$$

and $h(x)$ is locally integrable and defined by this equation for almost all x.

Proof.

$$\langle f * g, \varphi) = \langle f_\xi \otimes g_\eta, \varphi(\xi + \eta)\rangle = \iint f(\xi)\, g(\eta)\, \varphi(\xi + \eta)\, \mathrm{d}\xi\, \mathrm{d}\eta.$$

By making the change of variable $\xi = t$, $\eta = x - t$, (Jacobian = 1) and applying the Fubini theorem, we get

$$\langle f * g, \varphi\rangle = \iint f(t)\, g(x - t)\, \varphi(x)\, \mathrm{d}x\, \mathrm{d}t = \int \varphi(x)\, \mathrm{d}x \int f(t)\, g(x - t)\, \mathrm{d}t.$$

Let

$$h(x) = \int f(t)\, g(x - t)\, \mathrm{d}t,$$

then

$$\langle f * g, \varphi\rangle = \langle h, \varphi\rangle.$$

Theorem. In a Lie group V, $S * T$ has a meaning (i.e., the supports permit convolution), if either S or T has a compact support.

Proof. We must show that the map $(\xi, \eta) \to \xi\eta$ is proper for $\xi \in A$, $\eta \in B$. Suppose that S has a compact support A and T has the support B. Let D be a compact subset of V. The set $H = \{(\xi, \eta) : \xi \in A, \eta \in B, \xi\eta \in D\}$ remains compact because ξ is in the compact set A and

$\eta = \xi^{-1}(\xi\eta)$ for $\xi \in A$, $\xi\eta \in D$ remains in $A^{-1}D$. It follows that H is compact, therefore, the map is proper.

Remark. Similarly, $R * S * T$ has a meaning if two of them have a compact support.

Example. Consider $Y * \delta' * 1$, where Y is the Heaviside function:

$$Y(x) = \begin{cases} 0 & x < 0. \\ 1 & x > 0. \end{cases}$$

The supports of Y, δ' and 1, respectively, are $[0, \infty]$, $\{0\}$ and R. The map $(\xi, \eta, \zeta) \rightarrow \xi + \eta + \zeta$ is not proper, therefore the internal convolution is not allowed. However, $(Y * \delta') * 1$ and $Y * (\delta' * 1)$ each have a meaning, but they are not equal, i.e., the convolution is not associative.

$$(Y * \delta') * 1 = \delta * 1 = 1$$
$$Y * (\delta' * 1) = Y * 0 = 0.$$

Invariance under the Group of Translations

In this section we shall try to find the Hilbert spaces $\mathscr{H}$ which remain invariant under the action of a given group and, in particular, under the action of the group of translations.

Let G be a group of topological and algebraic automorphisms of the locally convex topological vector space E. As previously, $\mathfrak{H}$ denotes the space of Hilbert spaces $\mathscr{H}$ with continuous injection in E and $\overline{\mathscr{L}}_+(E', E)$ denotes the closed convex cone of positive anti-kernels from E' into E. We have seen that $\mathfrak{H} \approx \mathscr{L}_+(E', E)$.

An element $\sigma \in G$ defines a bijection from $\mathscr{H} \in \mathfrak{H}$ to the Hilbert space $\sigma\mathscr{H} \in \mathfrak{H}$ as follows:

$$\sigma: \mathscr{H} \rightarrow \sigma\mathscr{H} = \{\sigma h \in \mathscr{H}\}$$

where $\sigma\mathscr{H}$ has the transported Hilbert structure of $\mathscr{H}$:

$$(\sigma h/\sigma k)_{\sigma\mathscr{H}} = (h/k)_{\mathscr{H}}$$

Since σ operates on E, it also operates on the dual E'. In fact,

$$\langle \sigma e', \sigma e \rangle = \langle e', e \rangle \qquad e \in E, e' \in E'$$

and the automorphism $e' \rightarrow \sigma e'$ on E' is given by

$$\langle \sigma e', f \rangle = \langle e', \sigma^{-1} f \rangle, \qquad f \in E, \qquad e' \in E'.$$

Therefore, σ also operates on $\mathscr{L}(E', E)$. In fact

$$\sigma L \, . \, \sigma e' = \sigma(Le')$$

and the automorphism $L \to \sigma L$ of $\mathscr{L}(E', E)$ is given by

$$\sigma L \, . f' = \sigma L(\sigma^{-1} f').$$

Clearly, if L is positive, σL is also positive. By transport of structure we may easily prove the following theorem:

Theorem. If $\mathscr{H} \in \mathfrak{H}$ corresponds to $L \in \mathscr{L}_+(E', E)$, then $\sigma\mathscr{H}$ corresponds to σL for every $\sigma \in G$.

Thus, if $\mathscr{H} = \sigma \mathscr{H}$ for all $\sigma \in G$, then $L = \sigma L$ for all $\sigma \in G$, and in order to find the $\mathscr{H}$ which are invariant under G we may look for the L which are invariant under G.

Let us turn now to the case of vector-valued particles. Let G be a group of automorphisms σ which operate on the universe V and on the finite-dimensional vector space $\mathbf{F}$. Then σ operates also on $\mathscr{D}'(V)$ and hence it operates on $\mathscr{H} \subset \mathscr{D}'(V) \otimes \mathbf{F}$.

On the other hand, since σ operates on V it also operates on $V \times V$ ($\sigma(v, w) = (\sigma v, \sigma w)$) and on $\mathscr{D}'(V \times V)$, and since σ operates on $\mathbf{F}$ it also operates on $\overline{\mathbf{F}}$ ($\sigma\bar{f} = \overline{\sigma f}$). Hence σ operates on the anti-kernels $K_{x,\xi} \in \mathscr{D}'(V \times V) \otimes \mathbf{F} \otimes \overline{\mathbf{F}}$. In terms of a basis $\{f_i\}$ in $\mathbf{F}$ we have $K = \sum K_{ij} f_i \otimes \bar{f}_j$ with $K_{ij} \in \mathscr{D}'(V \times V)$ and then

$$\sigma K = \sum \sigma K_{ij} \, \sigma f_i \otimes \overline{\sigma f_j}$$

where

$$\langle \sigma K_{ij}, \varphi(x, \xi) \rangle = \langle K_{ij}, \varphi(\sigma x, \sigma\xi) \rangle.$$

Again, by transport of structure, it may be shown that if $\mathscr{H} \subset \mathscr{D}'(V) \otimes \mathbf{F}$ corresponds to $K \in \mathscr{D}'(V \times V) \otimes F \otimes \overline{\mathbf{F}}$ then $\sigma\mathscr{H}$ corresponds to σK for every $\sigma \in G$, and instead of looking for the $\mathscr{H}$ which are invariant under G we may look for the super-positive K invariant under G.

We shall look now for the anti-kernels $K_{x,\xi} \in \mathscr{D}'(V \times V) \otimes \mathbf{F} \otimes \overline{\mathbf{F}}$ which are invariant under the group of translations for $V = E_4$. Let the universe V be the affine space E_4 (or E) and the group of translation G, isomorphic to the vector space $\mathbf{E}_4$ (or $\mathbf{E}$) associated with the affine space E_4 (or E). *We assume* that an element of G (which operates on E as an affine operator) operates identically on the finite-dimensional vector space $\mathbf{F}$.

Suppose, first, that we have a function $K(x, \xi)$. In order that $K(x, \xi)$ be invariant under any translation we must have

$K(x - a, \xi - a) = K(x, \xi)$ for all a.

In particular, if we let $a = \xi$, then

$$K(x, \xi) = K(x - \xi, 0) = H(x - \xi)$$

where H is a function of the single variable $x - \xi$. Thus, in this case, in order to obtain a K invariant under translations we may start with a function H of one variable and replace this variable by $x - \xi$.

Let us generalize this notion to distributions $K_{x,\xi}$ on $E \times E$. Consider the C^∞ isomorphism

$$E \times E \leftarrow E \times \mathbf{E}$$

given by $(x, \xi) \leftarrow (x, \mathbf{u})$, where $\mathbf{u} = \mathbf{x} - \xi$, $\xi = x - \mathbf{u}$. By transport of structure it defines an isomorphism on the distributions

$$K_{x,\xi} \in \mathscr{D}'(E \times E) \leftrightarrow H_{x,\mathbf{u}} \in \mathscr{D}'(E \times \mathbf{E})$$

given by

$$\langle H_{x,\mathbf{u}}, \psi(x, \mathbf{u})\rangle = \langle K_{x,\xi}, \psi(x, \overrightarrow{x - \xi})\rangle, \psi \in \mathscr{D}(E, \mathbf{E})$$

and

$$\langle K_{x,\xi}, \varphi(x, \xi)\rangle = \langle H_{x,\overline{\mathbf{u}}}, \varphi(x, x - \mathbf{u})\rangle \quad \varphi \in \mathscr{D}(E, E).$$

For convenience, we shall use the following notation: given $K_{x,\xi}$, then $H_{x,\mathbf{u}} = K_{x,x-\mathbf{u}}$, and given $H_{x,\mathbf{u}}$, then $K_{x,\xi} = H_{x,x-\xi}$. In order to find the corresponding property of H when K is invariant under translations, consider the following commutative diagram:

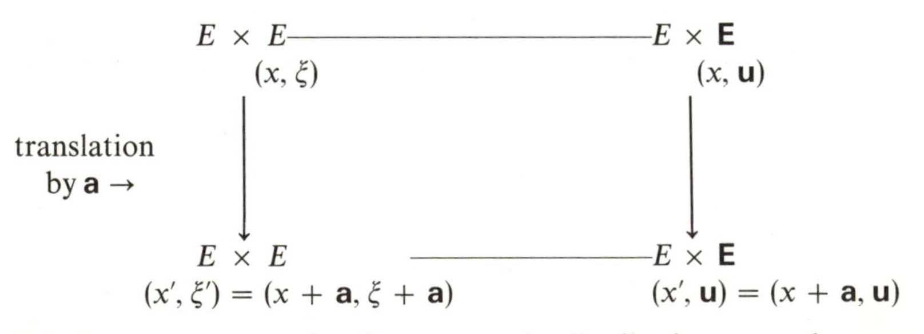

It induces a commutative diagram on the distributions over the corresponding spaces

where σ is a translation acting on the two variables x and ξ of K, and τ is the corresponding translation acting only on the first variable x of H. Thus, instead of looking for the $K \in \mathscr{D}'(E \times E) \otimes F \otimes \overline{F}$ which are invariant under all translations acting on two variables, we shall look for the $H \in \mathscr{D}'(E \times \mathbf{E}) \otimes F \otimes \overline{F}$ which are invariant under all translations acting on the first variable only.

Now, using the theorem of kernels:

$$\mathscr{D}'(E \times \mathbf{E}) \approx \mathscr{L}_b(\mathscr{D}(E), \mathscr{D}'(\mathbf{E})) = \mathscr{D}'(E; \mathscr{D}'(\mathbf{E}))$$

we see that

$$\mathscr{D}'(E \times \mathbf{E}) \otimes F \otimes \overline{F} \approx \mathscr{D}'(E; \mathscr{D}'(\mathbf{E}) \otimes F \otimes \overline{F}).$$

Thus $H_{x,\mathrm{u}}$ is a distribution on E having values in the infinite-dimensional locally convex topological vector space $\mathscr{F} = \mathscr{D}'(\mathbf{E}) \otimes F \otimes \overline{F}$, and we must search for those H which are invariant under all translations acting on E only. We write $H_{x,\mathrm{u}} \in \mathscr{D}'(E, \mathscr{F})$.

Let us choose a Lebesgue measure $\mathrm{d}x$ on E. The answer to our search is given by the following theorem.

Theorem. Every distribution $T \in \mathscr{D}'(E, \mathscr{F})$ which is invariant under all translations acting on E is a constant, i.e., $T = 1_{(x)} \otimes \boldsymbol{\lambda}$ where $\boldsymbol{\lambda} \in \mathscr{F}$ (or $T = \mathrm{d}x \otimes \boldsymbol{\lambda}$).

Proof. Let $\alpha \in \mathscr{D}(E)$ be an infinitely differentiable function such that $\int \alpha(x)\, \mathrm{d}x = 1$. Then for every $\varphi \in \mathscr{D}(E)$,

$$\langle T * \alpha, \varphi \rangle = \langle T_\xi \otimes \alpha(\eta), \varphi(\xi + \eta) \rangle = \int \alpha(\eta)\, d\eta \, \langle T_\xi, \varphi(\xi + \eta) \rangle$$
$$= \int \alpha(\eta)\, \mathrm{d}\eta \, \langle T_t, \varphi(t) \rangle = \langle T, \varphi \rangle$$

where the invariance of T under translation was used. Thus $T * \alpha = T$. According to a theorem given in the previous section (in the scaler case), T is an infinitely differentiable function, and since T is invariant under all translations it follows that T is a constant.

In our case if $H_{x,\mathrm{u}}$ is invariant under all translations on $x \in E$, then

$$H_{x,\mathrm{u}} = 1_x \otimes H_\mathrm{u} \quad (\text{or} \quad H_{x,\mathrm{u}} = \mathrm{d}x \otimes H_\mathrm{u})$$

where $H_\mathrm{u} \in \mathscr{F} = \mathscr{D}'(\mathbf{E}) \otimes F \otimes \overline{F}$.

We have shown up to now that the anti-kernels $K_{x,\xi} \in \mathscr{D}'(E \times E) \otimes F \otimes \overline{F}$, which are invariant under the group $\mathbf{E}$ of translations acting on $E \times E$, are in one-to-one correspondence with the distribution $H_{\mathbf{u}} \in \mathscr{D}'(\mathbf{E}) \otimes F \otimes \overline{F}$ on E with values in $F \otimes \overline{F}$. If $H_{\mathbf{u}}$ is given, then

$$K_{x,\xi} = H_{x,\overrightarrow{x-\xi}} = 1_{(x)} \otimes H_{\overrightarrow{x-\xi}} = H_{\overrightarrow{x-\xi}}$$

where, for every $\varphi \in \mathscr{D}(E \times E)$,

$$\langle K_{x,\xi}, \varphi(x, \xi)\rangle = \langle 1_{(x)} \otimes H_{\mathbf{u}}, \varphi(x, x - \mathbf{u})\rangle = \langle H_{\mathbf{u}}, \int \varphi(x, x - \mathbf{u})\, \mathrm{d}x\rangle$$
$$= \int \mathrm{d}x\, \langle H_{\mathbf{u}}, \varphi(x, x - \mathbf{u})\rangle.$$

(In computing $\langle H_{\mathbf{u}}, \varphi(x, x - u)\rangle$, x is considered fixed.)†

Let us now look for the conditions on $H_{\mathbf{u}}$ which correspond to positivity and super-positivity of $K_{x,\xi}$. Recall that $K_{x,\xi}$ is called positive if

$$\langle K_{x,\xi}, \varphi(x) \otimes \overline{\varphi(\xi)}\rangle \geqslant 0 \text{ in } F \otimes \overline{F}$$

for every $\varphi \in \mathscr{D}(E)$. In terms of $H_{\mathbf{u}}$ this condition becomes

$$\langle H_{\mathbf{u}}, \int \varphi(x)\overline{\varphi}(x - \mathbf{u})\, \mathrm{d}x\rangle \geqslant 0 \text{ in } F \otimes \overline{F}$$

for every $\varphi \in \mathscr{D}(E)$. In order to eliminate the presence of the affine space E, we may choose an origin in E, and if we recall that there is a one-to-one correspondence between E and $\mathbf{E}$, the condition of positivity becomes

$$\langle H_{\mathbf{u}}, \int \varphi(\mathbf{x})\, \overline{\varphi}(\mathbf{x} - \mathbf{u})\, \mathrm{d}\mathbf{x}\rangle \geqslant 0 \text{ in } F \otimes \overline{F}$$

for every $\varphi \in \mathscr{D}(\mathbf{E})$. The integral in this inequality would be a convolution if $\mathbf{x} - \mathbf{u}$ were replaced by $\mathbf{u} - \mathbf{x}$. In fact, if we use the notation

$$\check{\psi}(x) = \psi(-x)$$
$$\tilde{\psi}(x) = \check{\overline{\psi}}(x) = \overline{\psi}(-x)$$

the condition on H corresponding to positivity for K is that

$$\langle H, \varphi * \tilde{\varphi}\rangle \geqslant 0 \text{ in } F \otimes \overline{F}$$

for every $\varphi \in \mathscr{D}(\mathbf{E})$. Similarly, the condition on H corresponding to super-positivity for K is

$$\langle H, \psi * \tilde{\psi}\rangle \geqslant 0 \text{ in } C$$

for every $\psi \in \mathscr{D}(\mathbf{E}; F')$.

† Convolution of two vector—valued distributions can easily be defined, and has the same properties, provided one of them has values in a finite dimensional vector space.

The results of this section are summarized in the following theorem:

Theorem. Let E be a (finite-dimensional) affine space, and F a finite-dimensional vector space. The Hilbert spaces $\mathscr{H} \subset \mathscr{D}'(E:F)$ or the corresponding super-positive anti-kernels $K_{x,\xi} \in \mathscr{D}'(E \times E) \otimes F \otimes \overline{F}$, which are invariant under the group of translations $\mathbf{E}$ acting as affine operators on E and identically on F, are in one-to-one correspondence with the distributions $H_{\mathbf{u}} \in \mathscr{D}'(\mathbf{E}) \otimes F \otimes \overline{F}$ of super positive type. The condition of positivity is

$\langle H, \varphi * \tilde{\varphi} \rangle \geqslant 0$ in $F \otimes \overline{F}$ for every $\varphi \in \mathscr{D}(\mathbf{E})$, and the condition of super-positivity is

$\langle H, \varphi * \tilde{\varphi} \rangle \geqslant 0$ in C for every $\varphi \in \mathscr{D}(\mathbf{E}; F')$.

The above notions of positivity in $\mathscr{D}'(\mathbf{E}) \otimes F \otimes \overline{F}$ are natural extensions of the notion of positivity of functions in the sense of Bochner:

Definition. A function f defined on the l-dimensional vector space $\mathbf{E}$ is called *positive in the sense of Bochner* if, for every set of elements $\{x_1, \ldots, x_l\}$ in $\mathbf{E}$ and every set of complex numbers $\{z_1, \ldots, z_l\}$ in C,

$$\sum_{ij} f(x_1 - x_j)\, z_i \bar{z}_j \geqslant 0.$$

In the case of scalar-valued anti-kernels $K_{x,y}$ which are actually continuous functions $K(x, y)$ we saw that the classical definition of positivity coincides with the definition of positivity of anti-kernels ($\langle K, \varphi \otimes \overline{\varphi} \rangle \geqslant 0, \varphi \in \mathscr{D}$). Similarly, in the present case, if f is continuous, the condition of positivity in the sense of Bochner is equivalent to

$$\iint f(x - \xi)\, \varphi(x)\, \overline{\varphi}(\xi)\, \mathrm{d}x\, \mathrm{d}\xi \geqslant 0$$

or

$$\int f(u)\, \mathrm{d}u \int \varphi(x) \overline{\varphi}(x - u)\, \mathrm{d}x \geqslant 0$$

for every $\varphi \in \mathscr{D}$. Noting that if $H_{\mathbf{u}} \in \mathscr{D}'(\mathbf{E}) \otimes F \otimes \overline{F}$ is a continuous function $H(\mathbf{u})$, then $H_{\overrightarrow{x \xi}} = H(x - \xi)$, we may extend the notion of Bochner positivity to the space $\mathscr{D}'(\mathbf{E}) \otimes F \otimes \overline{F}$. Unfortunately, two definitions are possible:

Definition. A distribution $H \in \mathscr{D}'(\mathbf{E}) \otimes F \otimes \overline{F}$ is called

(1) *Positive in the sense of Bochner*, or *B-positive*, if

$$\langle H, \varphi * \tilde{\varphi} \rangle \geqslant 0 \text{ in } F \otimes \overline{F}$$

for every $\varphi \in \mathscr{D}(E)$.

(2) *Super-positive in the sense of Bochner*, or *B-super-positive*, if

$$\langle H, \psi * \tilde{\psi} \rangle \geqslant 0 \text{ in } C$$

for every $\psi \in \mathscr{D}(\mathbf{E}, F')$.

We shall discuss the equivalence of these definitions in a later section (*Bochner Theorem*).

Fourier Transforms

Let $\mathbf{E}$ be a finite-dimensional real vector space with a Lebesgue measure $d\mathbf{x}$. The *Fourier transform* $\mathscr{F} f$ of a function $f(\mathbf{x})$, $\mathbf{x} \in \mathbf{E}$, is given by the equation

$$\mathscr{F} f = g(\mathbf{p}) = \int \exp\left[-2i\pi\langle \mathbf{x}, \mathbf{p}\rangle\right] f(\mathbf{x})\, d\mathbf{x} \tag{3.1}$$

where $\mathbf{p} \in \mathbf{E}'$, the dual space of $\mathbf{E}$, and $\langle \mathbf{x}, \mathbf{p} \rangle$ is the duality product. A basis $\mathbf{e}_1, \mathbf{e}_2, \ldots, \mathbf{e}_n$ can be chosen in $\mathbf{E}$ in such a way that the Lebesgue measure is $dx_1 dx_2 \ldots dx_n$. The vector $p = \sum_i x_i \mathbf{e}_i$ may be represented by its coordinates $\{x_1\}$ and the vector $\mathbf{p} = \sum_i p_i \mathbf{e}'_i$ may be represented by its coordinates $\{p_i\}$. Equation (3.1) then becomes

$$\mathscr{F} f = g(p_1, p_2, \ldots, p_n) = \int \exp\left[-2i\pi(x_1 p_1 + x_2 p_2 + \ldots + x_n p_n)\right] f(x_1, x_2, \ldots, x_n)\, dx_1 dx_2 \ldots dx_n$$

When there is an inner product $(\mathbf{x} | \mathbf{y})$ given in $\mathbf{E}$, the Fourier transform may be defined by

$$\gamma(\mathbf{y}) = \int \exp\left[-2i\pi\,(\mathbf{x} | \mathbf{y})\right] f(\mathbf{x})\, d\mathbf{x};$$

however, we shall continue to use the definition of equation (3.1).

If $f \in L^1$, the integral (3.1) exists and $g(\mathbf{p})$ is continuous and bounded, in fact

$$|g(\mathbf{p})| \leqslant \|f\|_1.$$

If f has a Fourier image $\mathscr{F} f = g$ and $x_i f$ is integrable for each $j = 1, 2, \ldots, n$, then the derivatives $\partial g / \partial p_j$ exist, are continuous and bounded and

$$\mathscr{F}(-2i\pi\, x_j f) = \frac{\partial g}{\partial p_j}.$$

If f decreases rapidly at ∞, i.e., if for every polynomial P, the product Pf is integrable, then all derivatives of g exist and are bounded.

Let us integrate equation (3.1) by parts. We find

$$g(\mathbf{p}) = \int \mathrm{d}x_1 \ldots \mathrm{d}x_{j-1}\, \mathrm{d}x_{j+1} \ldots \mathrm{d}x_n$$

$$\left\{ \left[\frac{1}{-2i\,\pi p_j} \exp\left(-2i\pi\langle \mathbf{x}, \mathbf{p}\rangle\right) f \right]_{-\infty}^{\infty} - \int \frac{1}{-2i\,\pi p_j} \exp\left(-2i\pi\langle \mathbf{x}, \mathbf{p}\rangle\right) \frac{\partial f}{\partial x_j} \mathrm{d}x_j \right\}$$

Suppose that $f \in L^1$ and that the first derivatives $\partial f/\partial x_j$ are continuous and integrable; then it is easily seen that the first term vanishes. It follows that

$$2i\pi\, p_j\, g(\mathbf{p}) = \int \exp\left(-2i\,\pi\langle \mathbf{x}, \mathbf{p}\rangle\right) \frac{\partial f}{\partial x_j} \mathrm{d}\mathbf{x} \tag{3.2}$$

and $p_j\, g(\mathbf{p})$ is continuous and bounded.

If f is infinitely differentiable and is L^1 as well as all its derivatives, then the product of g with every polynomial remains continuous and bounded.

Definition. The space $\mathscr{S}(\mathbf{E})$ is the space of all infinitely differentiable functions with each derivative rapidly decreasing at ∞, i.e., for arbitrary polynomials P and Q, $P(x)Q(\partial/\partial x)f$ exists and is bounded, or (equivalently) $Q(\partial/\partial x)\,P(x)f$ exists and is bounded.

The following is easily proved from the previous results.

Theorem. $\varphi \in \mathscr{S}(\mathbf{E}) \Rightarrow \mathscr{F}\,\varphi \in \quad (\mathbf{E}')$.

Now Fourier transforms will be extended to distributions. First of all, note that if a Lebesgue measure is given in $\mathbf{E}$, then a Lebesgue measure in $\mathbf{E}'$ is also given, because if a basis in $\mathbf{E}$ is chosen such that with respect to this basis the measure is $\mathrm{d}x_1\, \mathrm{d}x_2 \ldots \mathrm{d}x_n$, then the measure with respect to the dual basis in $\mathbf{E}'$ is $\mathrm{d}p_1\, \mathrm{d}p_2 \ldots \mathrm{d}p_n$.

For $f \in L^1$, $\varphi \in \mathscr{D}_p(\mathbf{E}')$ let us take the duality product of φ with the Fourier transform $g(\mathbf{p}) = \mathscr{F} f$ of f:

$$\begin{aligned}\langle g, \varphi\rangle = \langle \mathscr{F} f, \varphi\rangle &= \int \varphi(\mathbf{p})\, \mathrm{d}\mathbf{p} \int \exp\left(-2i\pi\langle \mathbf{x}, \mathbf{p}\rangle\right) f(\mathbf{x})\, \mathrm{d}\mathbf{x} \\ &= \int f(\mathbf{x})\, \mathrm{d}\mathbf{x} \int \exp\left(-2i\pi\langle \mathbf{x}, \mathbf{p}\rangle\right) \varphi(\mathbf{p})\, \mathrm{d}\mathbf{p} = \langle f, \psi\rangle\end{aligned}$$

where

$$\psi = \int \exp\left(-2i\pi\langle \mathbf{x}, \mathbf{p}\rangle\right) \varphi(\mathbf{p})\, \mathrm{d}\mathbf{p} = \mathscr{F}\varphi.$$

Thus we have

$$\langle \mathscr{F} f, \varphi\rangle = \langle f, \mathscr{F}\varphi\rangle.$$

In view of this, we may try to define the Fourier transform of distributions as follows:

$$\langle \mathscr{F} T, \varphi \rangle = \langle T, \mathscr{F} \varphi \rangle$$

where $\varphi \in \mathscr{D}_p$, $T \in \mathscr{D}'_x$. We would like to have $\mathscr{F} \varphi \in \mathscr{D}_x$, but unfortunately $\mathscr{F} \varphi \notin \mathscr{D}_x$ unless $\varphi \equiv 0$.

At this time the space $\mathscr{S}(\mathbf{E})$ becomes useful. A topology on $\mathscr{S}$ is given by the following notion of convergence:

Definition. The functions $f_j \to 0$ in the sense of $\mathscr{S}$ if $P(x)\, Q(\partial/\partial x) f_j$ converges uniformly to zero for arbitrary polynomials P and Q.

The space $\mathscr{S}'(\mathbf{E})$ of continuous linear functionals over $\mathscr{S}(\mathbf{E})$ is called the space of tempered distributions.
Note that

$$\mathscr{D}(\mathbf{E}) \subset \mathscr{S}(\mathbf{E})$$

and

$$f_j \to 0 \text{ in } \mathscr{D} \Rightarrow f_j \to 0 \text{ in } \mathscr{S},$$

thus a continuous linear functional over $\mathscr{S}$ is a continuous linear functional over $\mathscr{D}$. Therefore there is a continuous linear map of $\mathscr{S}'(\mathbf{E})$ onto a subspace of $\mathscr{D}'(\mathbf{E})$. In fact this map is an injection, since $\mathscr{D}(\mathbf{E})$ is dense in $\mathscr{S}(\mathbf{E})$, as it can be proved easily.

Examples.

(1) $\varphi(x) = \exp[-x^2] \in \mathscr{S}(R)$ where (R) is the set of real numbers.

(2) Any function f, locally integrable and measurable such that $|f(x)| \leqslant P(x)$ for some polynomial $P(x)$ is a tempered distribution.

(3) e^x is not a tempered distribution.

Proof. Take $\varphi(x) = \exp[-\sqrt[4]{(x^2+1)}]$, then $\langle \exp x, \varphi \rangle =$
$= \int \exp x \exp[-\sqrt[4]{(x^2+1)}]\, dx$ does not converge.

Definition. For $T \in \mathscr{S}'_x$, the Fourier transform $\mathscr{F} T$ is defined by the equation

$$\langle \mathscr{F} T, \varphi \rangle = \langle T, \mathscr{F} \varphi \rangle$$

where $\varphi \in \mathscr{S}_p$.

Since $\varphi \in \mathscr{S}_p \Rightarrow \mathscr{F} \varphi \in \mathscr{S}_x$, the formula has a meaning for all $T \in \mathscr{S}'_x$. If $\varphi \to 0$ in $\mathscr{S}_p$, then $\mathscr{F} \varphi \to 0$ in $\mathscr{S}_x$ and

$$\langle \mathscr{F} T, \varphi \rangle = \langle T, \mathscr{F} \varphi \rangle \to 0,$$

thus $\mathscr{F} T$ is a continuous linear functional and $\mathscr{F} T \in \mathscr{S}'_p$.

Remark. If a distribution $T = f \in L^1$, then T is tempered and the Fourier transform of T given by the above definition is equal to the

Fourier transform of the function f given by equation (3.1):

$$\mathscr{F}T = \mathscr{F}f.$$

Some of the main properties of Fourier transforms of tempered distributions are now stated with the proofs of most of them omitted.

(1) $\mathscr{F}\delta = 1$

Proof. $\langle \mathscr{F}\delta, \varphi \rangle = \langle \delta, \mathscr{F}\varphi \rangle = \langle \delta, \int \exp[-2i\pi\langle \mathbf{x}, \mathbf{p} \rangle] \varphi(\mathbf{p}) \, d\mathbf{p} \rangle$

$$= \int \varphi(\mathbf{p}) \, d\mathbf{p} = \langle 1, \varphi \rangle$$

(2) $\mathscr{F}(\partial\delta/\partial x_j) = 2i\pi p_j$

and similarly for polynomials P,

$$\mathscr{F}P(\partial/\partial x)\,\delta = P(2i\pi p).$$

(3) *Reciprocity formula.* If $\overline{\mathscr{F}}$ is defined by replacing i by $-i$ in the definition of $\mathscr{F}$, the following is true:

$$\mathscr{F}\,\overline{\mathscr{F}} = \overline{\mathscr{F}}\,\mathscr{F} = I$$

$V = \mathscr{F}U \Leftrightarrow U = \overline{\mathscr{F}}V$. (If either U or V is tempered, then the other is tempered and this statement has a meaning.)

(4) $\overline{\mathscr{F}}\delta = 1$

(5) $\mathscr{F}1 = \overline{\mathscr{F}}1 = \delta$

(6) $\mathscr{F}(-2i\pi x_j) = \partial\delta/\partial p_j$

and similarly for polynomials P,

$\mathscr{F}P(-2i\pi x) = P(\partial/\partial p)\,\delta.$

Remark. Suppose that $g = \mathscr{F}f$ for $f \in L^1$. Then g is continuous and bounded, but might not be integrable. In that case, the inverse Fourier transform $f(\mathbf{x}) = \int \exp[2i\pi\langle \mathbf{x}, \mathbf{p} \rangle] g(\mathbf{p}) \, d\mathbf{p}$ does not exist in the classical sense. However, in the sense of distributions, the inverse Fourier transform does exist.

(7) $f \in L^2 \Leftrightarrow \mathscr{F}f \in L^2$ (true in the sense of distributions)

(8) $\|\mathscr{F}f\|_{L^2} = \|f\|_{L^2}$

(9) If $f, g \in L^2$ and $F = \mathscr{F}f$, $G = \mathscr{F}g$, then

$$\int f(\mathbf{x})\overline{g(\mathbf{x})} \, d\mathbf{x} = \int F(\mathbf{p})\overline{G(\mathbf{p})} \, d\mathbf{p}$$

i.e., $\mathscr{F}$ is a unitary operator.

(10) If $T \in \mathscr{S}'$ and T has a compact support, then $V(\mathbf{p}) = \langle T_x, \exp[-2i\pi\langle \mathbf{x}, \mathbf{p}\rangle]\rangle$ is an infinitely differentiable function of $\mathbf{p}$ and $V(\mathbf{p}) = \mathscr{F}T$.

Proof. Proof of differentiability was discussed in the section on tensor products. Now we have

$$\langle \mathscr{F}T, \varphi\rangle = \langle T, \mathscr{F}\varphi\rangle = \langle T_x, \int \exp[-2i\pi\langle \mathbf{x}, \mathbf{p}\rangle]\varphi(\mathbf{p})d\mathbf{p}\rangle$$
$$= \langle T_x \otimes \varphi(\mathbf{p}), \exp[-2i\pi\langle \mathbf{x}, \mathbf{p}\rangle]\rangle$$
$$= \int \varphi(\mathbf{p})\,d\mathbf{p}\,\langle T_x, \exp[-2i\pi\langle \mathbf{x}, \mathbf{p}\rangle]\rangle = \langle V, \varphi\rangle.$$

If $\mathbf{p}$ is allowed to have complex values, $V(p)$ becomes a holomorphic function.

(11) If $S, T \in \mathscr{S}'$, if T has a compact support (thus $S * T$ has a meaning), and if $U = \mathscr{F}S$, $V = \mathscr{F}T$, then UV has a meaning, $UV \in \mathscr{S}'$ and $UV = \mathscr{F}(S * T)$. Thus Fourier transforms change convolutions into multiplications and multiplications into convolutions.

Bochner's Theorem

Bochner's Theorem for Scalar Distributions.

A distribution T on $\mathbf{E}_n$ is of positive type if and only if T is tempered and its Fourier transform is a tempered positive measure.

We recall that T is of positive type if $< T, \varphi * \tilde{\varphi} > \geqslant 0$ whatever be $\varphi \in \mathscr{D}(\mathbf{E}_n)$. A proof of Bochner's theorem can be found in Schwartz, *Théorie des Distributions*, II, Chap. VI.

Definition. Let $\mathbf{M}$ be a finite-dimensional ordered vector space. We say that a distribution $\mathbf{T} \in \mathscr{D}'(\mathbf{E}_n) \otimes \mathbf{M}$ is B-positive if $< \mathbf{T}, \varphi * \tilde{\varphi} >$ is a positive element of $\mathbf{M}$ whatever be $\varphi \in \mathscr{D}(\mathbf{E}_n)$.

Bochner's Theorem for Vector Distributions.

A distribution $\mathbf{T} \in \mathscr{D}'(\mathbf{E}_n) \otimes \mathbf{M}$ is B-positive if and only if $\mathbf{T}$ is tempered and its Fourier transform $\mathscr{F}\,\mathbf{T}(\in \mathscr{D}'(\mathbf{E}'_n) \otimes \mathbf{M})$ is a tempered positive measure. $\mathbf{M}$ has to be assumed finite-dimensional.

Proof of: $\mathbf{T}$ *B-Positive* $\rightarrow$ $\mathbf{T}$ *Tempered and* $\mathscr{F}$ $\mathbf{T}$ *is a Tempered Positive Measure*

We show first that $\mathbf{T}$ is tempered. Let Γ be the closed convex cone which defines the order in $\mathbf{M}$, Γ' the dual cone in $\mathbf{M}'$. A distribution $\mathbf{T}$ valued in $\mathbf{M}$ is a continuous linear map $\mathscr{D}(\mathbf{E}_n) \rightarrow \mathbf{M}$; any $\tilde{\mathbf{m}}' \in \Gamma'$

defines a map $\mathbf{M} \to C$. Therefore the map $\mathbf{T}$ followed by the map $\mathbf{m}'$ is a scalar-valued distribution $\mathbf{m}' \circ \mathbf{T}$:

$$\langle \mathbf{m}' \circ \mathbf{T}, \varphi \rangle = \mathbf{m}'(\langle \mathbf{T}, \varphi \rangle).$$

Since $\mathbf{T}$ is *B-positive*,

$$\langle \mathbf{m}' \circ \mathbf{T}, \varphi * \tilde{\varphi} \rangle = \mathbf{m}'(\langle \mathbf{T}, \varphi * \tilde{\varphi} \rangle) \geqslant 0$$

for every $\varphi \in \mathscr{D}(\mathbf{E})$, $\mathbf{m}' \in \Gamma'$. Hence $\mathbf{m}' \circ \mathbf{T}$ is a scalar distribution of positive type for every $\mathbf{m}' \in \Gamma'$. By Bochner's theorem for scalar distributions, $\mathbf{m}' \circ \mathbf{T}$ is a tempered distribution. Since finite linear combinations of tempered distributions are tempered, and since Γ' spans $\mathbf{M}'$, $\mathbf{m}' \circ \mathbf{T}$ is a tempered distribution whatever be $\mathbf{m}' \in \mathbf{M}'$, $\mathbf{T}$ is tempered.

To show that $\boldsymbol{\mu} = \mathscr{F}\,\mathbf{T}$ is a tempered positive measure, first note that, whatever be $\mathbf{m}' \in \Gamma'$, $\mathbf{m}' \circ \boldsymbol{\mu} = \mathbf{m}' \circ \mathscr{F}\,\mathbf{T} = \mathscr{F}(\mathbf{m}' \circ \mathbf{T})$ is a tempered positive measure according to Bochner's theorem for scalar distributions. Since Γ' spans $\mathbf{M}'$, $\mathbf{m}' \circ \mu$ is a tempered measure for every $\mathbf{m}' \in \mathbf{M}'$. Therefore, $\boldsymbol{\mu}$ is a tempered measure.

To show that $\boldsymbol{\mu}$ is positive, we need to show that $\langle \boldsymbol{\mu}, \varphi \rangle \in \Gamma$ for any continuous function $\varphi \geqslant 0$ with compact support. It is sufficient to show that $\mathbf{m}'(\langle \boldsymbol{\mu}, \varphi \rangle) \geqslant 0$ for arbitrary $\mathbf{m}' \in \Gamma'$. But since $\mathbf{m}' \circ \boldsymbol{\mu}$ is a positive (scalar) measure,

$$\mathbf{m}'(\langle \boldsymbol{\mu}, \varphi \rangle) = \langle \mathbf{m}' \circ \boldsymbol{\mu}, \varphi \rangle \geqslant 0.$$

Proof of: $\mathscr{F}\,\mathbf{T}$ *tempered positive measure* $\to$ $\mathbf{T}$ *B-positive*

Proof. Any $\psi \in \mathscr{S}(\mathbf{E}'_n)$, $\psi \geqslant 0$, is the limit in $\mathscr{S}$ of a sequence of functions $\alpha_\nu \psi$ where α_ν can be chosen as follows:

(1) $\alpha_\nu \in \mathscr{D}(\mathbf{E}'_n)$;

(2) $0 \leqslant \alpha_\nu \leqslant 1$;

(3) $\alpha_\nu - 1 \to 0$ uniformly as well as every derivative on every compact set.

Since $\alpha_\nu \psi$ is continuous with compact support and $\boldsymbol{\mu} = \mathscr{F}\mathbf{T}$ is a positive measure, we have

$$\langle \boldsymbol{\mu}, \alpha_\nu \psi \rangle \geqslant 0 \text{ in } \mathbf{M}.$$

By going to the limit (since $\boldsymbol{\mu}$ is tempered), we get:

$$\langle \boldsymbol{\mu}, \psi \rangle \geqslant 0 \text{ in } \mathbf{M}.$$

This holds true for any $\psi \in \mathscr{S}(\mathbf{E}')$, $\psi \geqslant 0$. Let $\psi = \varphi\tilde{\varphi}$, $\varphi \in \mathscr{S}$.

Then

$$0 \leqslant \langle \boldsymbol{\mu}, \varphi\overline{\varphi} \rangle = \langle \mathscr{F}\mathbf{T}, \varphi\overline{\varphi} \rangle$$
$$= \langle \mathbf{T}, \mathscr{F}(\varphi\overline{\varphi}) \rangle = \langle \mathbf{T}, \mathscr{F}\varphi * \mathscr{F}\overline{\varphi} \rangle$$

But $\mathscr{F}\varphi$ can be any function θ of $\mathscr{S}(\mathbf{E}_n)$ and $\mathscr{F}\overline{\varphi} = \tilde{\theta}$. Therefore for any $\theta \in \mathscr{S}$,

$$\langle \mathbf{T}, \theta * \tilde{\theta} \rangle \geqslant 0 \text{ in } \mathbf{M}.$$

Now we shall examine *B-superpositivity*.

Lemma 1. Let ψ be a continuous function with compact support, valued in a finite-dimensional ordered vector space $\mathbf{N}$, and $\psi \geqslant 0$. Then ψ can be uniformly approximated by finite sums of the form $\sum \alpha_i \psi_i$, where the ψ_i are vectors $\geqslant 0$ in $\mathbf{N}$ and the α_i are continuous non-negative scalar functions with compact support.

Proof. Let K be the support of ψ. Choose any $\varepsilon \geqslant 0$. Since K is compact and ψ is continuous, there is a finite open covering $\{\Omega_i\}$ of K such that the oscillation of ψ in each Ω_i is $\leqslant \varepsilon$. Let $\{\alpha_i\}$ be a partition of unity on K subordinated to the covering $\{\Omega_i\}$. Let ψ_i be the value of ψ at an arbitrary point of Ω_i. We have $\psi_i \geqslant 0$, and also:

$$\| \psi - \sum \alpha_i \psi_i \| = \| \sum \alpha_i \psi - \sum \alpha_i \psi_i \| \leqslant \sum \alpha_i \varepsilon = \varepsilon.$$

Lemma 2. If $\boldsymbol{\mu}$ is a positive measure on $\mathbf{E}'_n$ with values in a finite-dimensional ordered vector space $\mathbf{M}$, then $\langle \boldsymbol{\mu}, \psi \rangle \geqslant 0$ (in the sense of complex numbers) whatever be the continuous function ψ with compact support, valued and positive in $\mathbf{M}'$ (for the dual order relation).

Proof. Given any $\psi \in \mathscr{D}_1(\mathbf{E}'_n) \otimes \mathbf{M}$ such that $\psi \geqslant 0$, by Lemma 1 it can be approximated uniformly by finite sums of the form $\sum \mathbf{m}_i \alpha_i$, where $\mathbf{m}'_i \geqslant 0$ (in $\mathbf{M}'$), $\alpha_i \geqslant 0$ and α_i are continuous scalar functions with compact support. We have

$$\langle \boldsymbol{\mu}, \sum \mathbf{m}'_i \alpha_i \rangle = \sum \langle \langle \boldsymbol{\mu}, \alpha_i \rangle, \mathbf{m}'_i \rangle \geqslant 0$$

since

$$\langle \boldsymbol{\mu}, \alpha_i \rangle \geqslant 0 \text{ in } \mathbf{M} \text{ and } \tilde{\mathbf{m}}'_i \geqslant 0 \text{ in } \mathbf{M}'.$$

Let $\mathbf{F}$ be a finite-dimensional vector space.

Theorem. Let $\boldsymbol{\mu}$ be a tempered positive measure on $\mathbf{E}'$ with values in $F \otimes \overline{F}$ Then its Fourier transform $\mathbf{T} = \overline{\mathscr{F}}\boldsymbol{\mu}$ is a B-superpositive distribution.

In this statement, $\boldsymbol{\mu}$ is positive in the sense of the "natural" order relation in $F \otimes \overline{F}$.

Proof. Consider any $\varphi \in \mathscr{S}(\mathbf{E}_n') \otimes F'(F'$: dual of $F)$. Then $\varphi \otimes \overline{\varphi} \geqslant 0$ in $F' \otimes F'$. By Lemma 2, and the fact that φ is the limit in $\mathscr{S}(\mathbf{E}_n')$ of a sequence of functions in $\mathscr{D}(\mathbf{E}_n') \otimes \mathbf{F}'$, we have

$$\langle \mu, \varphi \otimes \overline{\varphi} \rangle \geqslant 0 \text{ in } C.$$

Then

$$\begin{aligned} 0 \leqslant \langle \mu, \varphi \otimes \overline{\varphi} \rangle &= \langle \mathscr{F}\mathbf{T}, \varphi \otimes \overline{\varphi} \rangle \\ &= \langle \mathbf{T}, \mathscr{F}(\varphi \otimes \overline{\varphi}) \rangle = \langle \mathbf{T}, \mathscr{F}\varphi * \mathscr{F}\overline{\varphi} \rangle \\ &= \langle \mathbf{T}, \theta * \tilde{\theta} \rangle \end{aligned}$$

where $\theta = \mathscr{F}\varphi$ can be any element of $\mathscr{S}(\mathbf{E}_n') \otimes \mathbf{F}'$.

Corollary. For any distribution $\mathbf{T} \in \mathscr{D}'(\mathbf{E}_n) \otimes F \otimes \overline{F}$, the two following properties are equivalent:

(1) $\mathbf{T}$ is B-positive;
(2) $\mathbf{T}$ is B-superpositive.

Summary and Reduction of the Problem

Let us briefly summarize the situation so far. We are given a C^∞-manifold, the so-called *universe* V, and a finite-dimensional vector space $\mathbf{F}$ over the complex numbers C. We are looking for $\mathbf{F}$-*particles* on V. Each is an Hilbert space $\mathscr{H} \subset \mathscr{D}'(V) \otimes \mathbf{F}$ (continuous injection). A *motion* of the particle is then a distribution $\psi \in \mathscr{H}$ such that $\| \psi \|_{\mathscr{H}} = 1$.

We have seen the following:

(1) There is a one-to-one correspondence between the $\mathbf{F}$-particles $\mathscr{H}$ and the positive anti-kernels, that is, the positive anti-linear continuous maps $\mathscr{D}(V) \otimes \mathbf{F}' \to \mathscr{D}'(V) \otimes \mathbf{F}$. This correspondence preserves the algebraic and order relations.

(2) The positive anti-kernels are in one-to-one correspondence with superpositive distributions $K_{x,\xi} \in \mathscr{D}'(V \otimes V) \otimes \mathbf{F} \otimes \overline{\mathbf{F}}$ ($\overline{\mathbf{F}}$ is a complex conjugate of $\mathbf{F}$), that is, distributions $K_{x,\xi}$ such that

$$\langle K_{x,\xi}, \varphi(x) \otimes \overline{\varphi(\xi)} \rangle \geqslant 0$$

whatever be $\varphi \in \mathscr{D}(V) \otimes \mathbf{F}'$.

If $K_{x,\xi}$ is given, the associated anti-kernel K is given by:

$$\langle K . \overline{\varphi}, \psi \rangle = \langle K, \psi(x) \otimes \overline{\varphi(\xi)} \rangle$$

whatever be φ, $\psi \in \mathscr{D}(V) \otimes \mathbf{F}'$; $\overline{\varphi}$ is the conjugate of φ in the sense of the conjugation $\mathbf{F} \to \overline{\mathbf{F}}$ or, more exactly here, its transposed $\overline{\mathbf{F}}' \to \mathbf{F}'$. Here again, the correspondence preserves the various algebraic and order relations.

If, moreover, we are given a structural group G, operating in a suitable sense on V and on $\mathbf{F}$, the space $\mathscr{H}$ will be G-invariant if the distribution $K_{x,\xi}$ is G-invariant.

(3) We place ourselves, then, in a particular case: $V = E_n$, an n-dimensional affine space, $G = \mathbf{E}_n$, the group of translations on E_n, operating identically on $\mathbf{F}$. Now consider the mapping

$$E_n \times E_n \to E_n \times \mathbf{E}_n, \qquad (x, \xi) \to (x, \mathbf{u})$$

where $\mathbf{u} = \overrightarrow{x - \xi}$, and the image $H_{x,\mathbf{u}}$ of $K_{x,\xi}$ under that mapping. Conversely, given

$$H_{x,\vec{\mathbf{u}}} \in \mathscr{D}'(E_n \times \mathbf{E}_n) \otimes \mathbf{F} \otimes \overline{\mathbf{F}},$$

wemay get $K_{x,\xi}$ by $K_{x,\xi} = H_{x,\overrightarrow{x-\xi}}$.

If $K_{x,\xi}$ is invariant under the (particular) translations $(x, \xi) \to (x + \boldsymbol{\alpha}, \xi + \boldsymbol{\alpha})$, $\boldsymbol{\alpha} \in \mathbf{E}_n$, then $H_{x,\mathbf{u}}$ must be invariant under the translations $(x, \mathbf{u}) \to (x + \boldsymbol{\alpha}, \mathbf{u})$. The latter invariance implies that $H_{x,\mathbf{u}} = 1_x \otimes H_{\mathbf{u}}$, where 1_x is the function 1 in the x-variable (that is, on E_n) and $H_{\mathbf{u}} \in \mathscr{D}'(\mathbf{E}_n) \otimes \mathbf{F} \otimes \overline{\mathbf{F}}$.

Super-positivity of $K_{x,\xi}$ is equivalent to B-superpositivity of $H_{\mathbf{u}}$. But we have seen that this is equivalent to B-positivity.

Now this means that $H_{\mathbf{u}}$ is tempered and its Fourier transform is a positive tempered measure μ with values in $\mathbf{F} \otimes \overline{\mathbf{F}}$. Conversely, to every such measure corresponds a distribution $H_{\mathbf{u}}$ B-positive.

Remark. Let K be the anti-kernel corresponding to the particle $\mathscr{H}$. Then $K \,.\, \overline{\boldsymbol{\varphi}} = H * \overline{\boldsymbol{\varphi}}$, $\boldsymbol{\varphi} \in \mathscr{D}(E_n) \otimes \mathbf{F}'$.

Proof. We have, for any $\boldsymbol{\psi} \in \mathscr{D}(E_n) \otimes \mathbf{F}'$,

$$\begin{aligned}\langle K \,.\, \overline{\boldsymbol{\varphi}}, \boldsymbol{\psi}\rangle &= \langle K_{x,\xi}, \boldsymbol{\psi}(x) \otimes \overline{\boldsymbol{\varphi}(\xi)}\rangle = \langle K_{x,\xi}, \boldsymbol{\psi}(x)\, \overline{\boldsymbol{\varphi}(\xi)}\rangle \\ &= \langle 1_x \otimes H_{\mathbf{u}}, \boldsymbol{\psi}(x)\, \overline{\boldsymbol{\varphi}}(x - \mathbf{u})\rangle = \int \boldsymbol{\psi}(x)\, \mathrm{d}x\, \langle H_{\mathbf{u}}, \overline{\boldsymbol{\varphi}}(x - \mathbf{u})\rangle \\ &= \langle (H * \overline{\boldsymbol{\varphi}})(x), \boldsymbol{\psi}(x)\rangle \,. \text{ Q.E.D.}\end{aligned}$$

Note that $H * \overline{\boldsymbol{\varphi}}$ is a convolution between a distribution on a vector space and a distribution on an affine space associated to the vector space. Let E be an affine space, $\mathbf{E}$ its associated vector space, S_x a distribution on E, $T_{\mathbf{x}}$ one on $\mathbf{E}$ (for instance, one at least of the two with compact support).

Then the general definition of $S_x * T_x$ is:

$$\langle S_x * T_x, \varphi(x)\rangle = \langle S_\xi \otimes T_\eta, \varphi(\xi + \eta)\rangle.$$

Then $S_\xi \otimes T_\eta$ is a distribution on the affine space E. We also have

$$\|H * \overline{\varphi}\| = \langle H * \overline{\varphi}, \varphi\rangle^{\frac{1}{2}} = \langle H, \varphi * \tilde{\varphi}\rangle^{\frac{1}{2}}$$

In the last term $\varphi * \tilde{\varphi}$ must be defined by choosing an origin because φ is defined on an affine space. Note that it is also necessary to choose an origin to carry out a Fourier transform of a distribution on an affine space.

Before stating the next theorem, let us observe that if E is a complete locally convex topological vector space, then to say that $\mathscr{H} \subset E$ *corresponds to the anti-kernel* L means that L is given by the following composition of maps

$$E' \to \mathscr{H}' \to \mathscr{H} \to E$$

where $\mathscr{H} \to E$ is the canonical embedding, $E' \to \mathscr{H}'$ its transposed map, and $\mathscr{H}' \to \mathscr{H}$ is the canonical anti-isomorphism of Hilbert spaces.

In the following, we shall have $\mathscr{H} \subset E_1 \subset E$, where all the injections are continuous. In particular, to $\mathscr{H}$ will correspond an anti-kernel L_1: $E_1' \to E_1$ and also an anti-kernel L: $E' \to E$. To distinguish the two we shall call L_1 an E_1-anti-kernel and L an E-anti-kernel.

Theorem. Let E, E_1 be two complete locally convex spaces, $E_1 \subset E$ with a finer topology than the one induced by E. Let $\mathscr{H} \subset E$ be a Hilbert space with a finer topology than the one induced by E, L the E-anti-kernel corresponding to $\mathscr{H}$.

Then, $\mathscr{H} \subset E_1$ with a finer topology than the one induced by E_1 if and only if L can be factorized as

$$E' \to E_1' \to E_1 \to E, \tag{3.3}$$

where $E_1 \to E$ is the natural injection, $E' \to E_1'$ its transposed map and $E_1' \to E_1$ is a continuous anti-linear map L_1. If L can be factorized in this way, L_1 is the E_1-anti-kernel corresponding to $\mathscr{H} \subset E_1$.

Proof. Let us look at the diagram

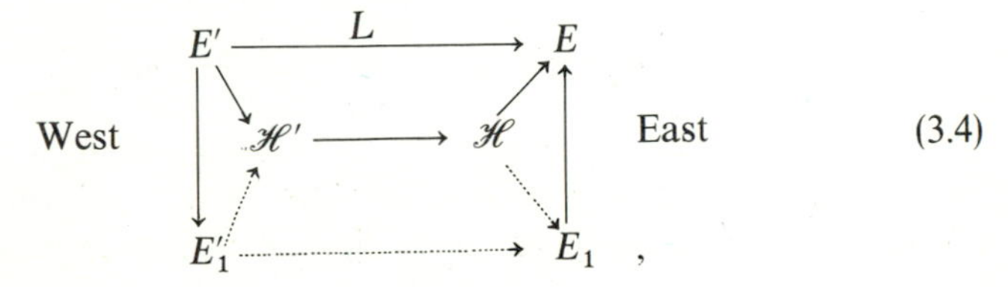

where $\mathscr{H} \to E$, $E_1 \to E$ are the natural embeddings, $E' \to \mathscr{H}'$, $E' \to E_1'$ the respective transposed maps, $\mathscr{H}' \to \mathscr{H}$ the canonical anti-isomorphism between $\mathscr{H}$ and its dual. Now if we assume $\mathscr{H} \subset E_1$ with continuous injection, we can then give a meaning to the dotted arrows: $\mathscr{H} \dashrightarrow E_1$ is the natural embedding, $E_1' \dashrightarrow \mathscr{H}'$ its transposed map, and L_1 is defined by the composition

$$E_1' \dashrightarrow \mathscr{H}' \to \mathscr{H} \dashrightarrow E_1$$

The factorization (3.3) follows from the commutativity of diagram (3.4) (that is, the fact that from West to East one gets identical results by following any path of arrows). Conversely, assume factorization (3.1). To the E_1-anti-kernel corresponds an Hilbert space $\mathscr{H}_1 \subset E_1$ with continuous injection. Then L_1 can be factorized as

$$E_1' \to \mathscr{H}_1' \to \mathscr{H}_1 \to E_1.$$

This implies that diagram (3.4), with $\mathscr{H}_1$ instead of $\mathscr{H}$ is commutative. Therefore, $\mathscr{H}_1$ is an Hilbert space continuously embedded in E corresponding to the E-anti-kernel L. But because of the one-to-one correspondence between E-anti-kernels and Hilbert spaces continuously embedded in E, we must have $\mathscr{H}_1 = \mathscr{H}$. Hence, here again we have a commutative diagram (3.4).

This theorem simplifies our search for the particles $\mathscr{H} \subset \mathscr{D}'(E) \otimes \mathbf{F}$. To $\mathscr{H}$ corresponds the map $K . \overline{\varphi} = H * \overline{\varphi}$:

$$\varphi \in \mathscr{D}(E) \otimes \mathbf{F}' \to H * \overline{\varphi} \in \mathscr{D}'(E) \otimes \mathbf{F},$$

but H is tempered, therefore $H * \overline{\varphi}$ not only defines a map from $\mathscr{D}(E) \otimes \mathbf{F}'$ to $\mathscr{D}'(E) \otimes \mathbf{F}$, but also defines a map from $\mathscr{S}(E) \otimes \mathbf{F}'$ to $\mathscr{S}(E) \otimes \mathbf{F}$. Thus the map $H * \overline{\varphi}$ can be factorized into the following

$$\mathscr{D}(E) \otimes \mathbf{F}' \to \mathscr{S}(E) \otimes \mathbf{F}' \to \mathscr{S}(E) \otimes \mathbf{F} \to \mathscr{D}'(E) \otimes \mathbf{F}.$$

According to the above theorem, if $\mathscr{H} \subset \mathscr{D}'(E) \otimes \mathbf{F}$ with a continuous injection, then $\mathscr{H} \subset \mathscr{S}'(E) \otimes \mathbf{F}$ with a continuous injection. The problem is thus reduced to the search for particles in $\mathscr{S}'(E) \otimes \mathbf{F}$.

The Space $\mathscr{H}$ for Scalar Particles

The case of scalar particles is defined by the fact that $\mathbf{F} = C$. Because of Bochner's theorem, we may now deal with tempered positive measures μ on $\mathbf{E}'$ instead of B-positive distributions $H \in \mathscr{D}'(\mathbf{E})$.

Given any such measure μ on $\mathbf{E}'$, we may consider L^2-functions with respect to μ. Each element $\dot{f}$ of L^2_μ is a class of μ—measurable functions f on $\mathbf{E}'$, μ—almost everywhere equal and such that

$$\int_{\mathbf{E}} \|f(p)\|^2 \, \mathrm{d}\mu(p) < +\infty.$$

The inner product in L^2_μ is:

$$(\dot{f}|\dot{g})_{L^2_\mu} = \int f\bar{g} \, \mathrm{d}\mu,$$

where f and g are any representatives of their classes $\dot{f}$ and $\dot{g}$, respectively.

We may associate to L^2_μ the space Λ^2_μ of distributions on $\mathbf{E}'$ of the form $f\mu$, where $f \in L^2_\mu$; note that then f is locally integrable with respect to μ. The distribution $f\mu$ is defined by the equation

$$\langle f\mu, \varphi \rangle = \int f\varphi \, \mathrm{d}\mu, \qquad \varphi \in \mathscr{D}(\mathbf{E}').$$

We see that $f\mu$ defines a new measure which is identical for all f in the same class $\dot{f}$. Set

$$(f\mu|g\mu)_{\Lambda^2_\mu} = \int f\bar{g} \, \mathrm{d}\mu.$$

This inner product turns Λ^2_μ into a Hilbert space; Λ^2_μ is continuously embedded in $\mathscr{D}'(\mathbf{E}')$.

Theorem. If μ is a tempered positive measure on $\mathbf{E}'$, then Λ^2_μ is continuously embedded in $\mathscr{S}'(\mathbf{E}')$.

Proof. For $\varphi \in \mathscr{D}, f \in L^2_\mu$ we have:

$$|\langle f\mu, \varphi \rangle| = |\int f\varphi \, \mathrm{d}\mu| \leqslant (\int |f|^2 \, \mathrm{d}\mu)^{\frac{1}{2}} (\int |\varphi|^2 \, \mathrm{d}\mu)^{\frac{1}{2}}$$

Since μ is tempered, any bounded set in $\mathscr{S}$ is bounded in L^2_μ. If $\varphi \in \mathscr{S}$, there is a sequence of elements φ_j of $\mathscr{D}$ converging to φ in $\mathscr{S}$, therefore remaining bounded in $\mathscr{S}$. This means that the above inequality remains true for $\varphi \in \mathscr{S}$ and if φ runs over a bounded set in $\mathscr{S}$ and f over the unit ball of L^2_μ, $|\langle f\mu, \varphi \rangle|$ remains bounded. This proves the theorem.

The space Λ^2_μ may be regarded as the completion of the vector space of distributions $f\mu$, where f runs over $\mathscr{S}$, for the norm $\|f\|_{\Lambda^2_\mu} = (\int |f|^2 \, \mathrm{d}\mu)^{\frac{1}{2}}$. We mean here the concrete completion in $\mathscr{S}'$.

On the other hand, the particle $\mathscr{H}$ is the completion in $\mathscr{D}'(\mathbf{E})$ of the space of elements $H * \varphi$, where φ runs over $\mathscr{S}$, for the norm $\|H * \varphi\|_{\mathscr{H}} = (\int |\mathscr{F}\varphi|^2 \, \mathrm{d}\mu)^{\frac{1}{2}}$. It means that *$\Lambda^2_\mu$ is the Fourier transform of $\mathscr{H}$*:

$$\Lambda^2_\mu = \mathscr{F}\mathscr{H}.$$

CHAPTER 4

Elementary Particles and the Lorentz Invariance

Elementary Particles

Let $V = E_n$, and let $\mathbf{F}$ be a finite-dimensional vector space over C. The structure group G operates on E_n and on $\mathbf{F}$. Assume that G satisfies the following properties in addition to the properties given in the definition of a structure group:

(1) G covers the whole group of translations; that is, for any translation $\boldsymbol{\alpha} \in \mathbf{E}_n$ there exists an element $g \in G$ such that the operation of g_{α} on E_n is the translation $\boldsymbol{\alpha}$ and g_{α} operates as the identity on $\mathbf{F}$.

(2) Every $\sigma \in G$ preserves Lebesgue measure on E_n. This is equivalent to saying that each σ has determinant ± 1 as an operator on $\mathbf{E}_n$.

Let $G_0 \subset G$ be the subset of G that operates as translations on $\mathbf{E}_n$ and as the identity on $\mathbf{F}$; G_0 is an invariant subgroup of G.

In order to see how σ operates on $K_{x,\xi}$, we make the usual "change of variables" $u = \overrightarrow{x - \xi}$, and write $K_{x,\xi} = H_{x,\mathbf{u}}$. It has already been shown that when $K_{x,\xi}$ is invariant under the transformation

$$(x, \xi) \to \sigma(x, \xi) = (\sigma x, \sigma\xi)$$

(with σ operating identically on $\mathbf{F}$) for $\sigma \in G_0$, then $H_{x,\mathbf{u}}$ is invariant under the transformation $(x, \mathbf{u}) \to (\sigma x, \mathbf{u})$.

It follows that $H_{x,\mathbf{u}} = 1_x \otimes H_{\mathbf{u}}$; note that $\sigma \in G_0$ operates identically on $H_{\mathbf{u}}$. Any $\sigma \in G$ operates on one hand on $\mathbf{F}$, on the other on E_n as $x \to \sigma x$, on $\mathbf{E}_n$ as $\mathbf{x} \to \boldsymbol{\sigma}\mathbf{x}$.

The operation $\boldsymbol{\sigma}$ is defined as follows: if $\xi, \eta \in E_n$ are such that $\overrightarrow{\xi - \eta} = \mathbf{x}$, then $\boldsymbol{\sigma}\mathbf{x} = \overrightarrow{\sigma\xi - \sigma\eta}$. This does not depend on the choice of $\xi, \eta \in E_n$; for let $\xi', \eta' \in E_n$ be such that $\overrightarrow{\xi' - \eta'} = \mathbf{x}$ also. Then $\overrightarrow{\xi - \xi'} = \overrightarrow{\eta - \eta'} = \boldsymbol{\alpha}$ or $\xi' = \xi - \alpha$, $\eta' = \eta - \boldsymbol{\alpha}$. Let us denote τ_{α} the translation $x \to x - \boldsymbol{\alpha}$ in E_n. We see that $\sigma\xi' = \sigma\tau_{\alpha}\xi$, $\sigma\eta' = \sigma\tau_{\alpha}\eta$. But there exists $\boldsymbol{\beta} \in \mathbf{E}_n$ such that $\sigma\tau_{\alpha} = \tau_{\beta}\sigma$; hence, for that $\boldsymbol{\beta}$, $\sigma\xi' =$

$\tau_\beta \sigma\xi = \sigma\xi - \boldsymbol{\beta}$, $\sigma\eta' = \tau_\alpha \sigma\eta = \sigma\eta - \boldsymbol{\beta}$ and $\sigma\xi' - \sigma\eta' = \sigma\xi - \sigma\eta$, which proves our remark. Then $\sigma \in G$ operates also on $E_n \times E_n$ as $(\xi, \eta) \to (\sigma\xi, \sigma\eta)$ and on $E_n \times \mathbf{E}_n$ as $(\xi, \mathbf{x}) \to (\sigma\xi, \boldsymbol{\sigma}\mathbf{x})$.

Now since σ preserves the Lebesgue measure on E_n, it preserves the constant functions (regarded as distributions) and therefore transforms $1_x \otimes H_\mathbf{u}$ into $1_x \otimes \sigma H_\mathbf{u}$. The invariance of $K_{x,\xi}$ under σ is equivalent to the invariance of $H_\mathbf{u}$ under σ (operating on $\mathbf{F}$ and on $\mathbf{E}_n$). Actually, we may replace invariance of $H_\mathbf{u}$ under G by its invariance under the factor group $G|G_0$. For G_0 operates identically on both $\mathbf{E}_n$ and $\mathbf{F}$.

For instance, if G is the inhomogeneous Lorentz group and G_0 is the group of translations, then $G|G_0$ is the homogeneous Lorentz group.

We want to find out the invariance properties of the Fourier transform $\mu = \mathscr{F}H$. In principle, $\sigma \in G$ operates on μ by operating on $\mathbf{F}$ on one hand and on $\mathbf{E}_n$ on the other as the contragredient operation of $\mathbf{u} \to \boldsymbol{\sigma}\mathbf{u}$, $\mathbf{u} \in \mathbf{E}$. Each $\mathbf{F}$-particle on E_n corresponds to a tempered positive measure μ on $\mathbf{E}_n'$ with values in $\mathbf{F} \otimes \overline{\mathbf{F}}$, invariant under $G|G_0$ (or G).

We go back to the general case, with the data of a universe V, a complex vector space $\mathbf{F}$, and a structural group G. Let $\mathscr{H}$ be a G-invariant $\mathbf{F}$-particle $\mathscr{H} \subset \mathscr{D}'(V) \otimes \mathbf{F}$, that is, a *universal* $\mathbf{F}$-particle, or also a G-$\mathbf{F}$-particle.

Definition. A universal $\mathbf{F}$-particle $\mathscr{H}$ is called *elementary* if any Hilbert space $\mathscr{H}_1 \leqslant \mathscr{H}$ (that is, $\mathscr{H}_1 \subset \mathscr{H}$ with a greater or equal norm) and G-invariant is either $\{0\}$ or $\mathscr{H}$ itself with a proportional norm.

Theorem. The three following properties of a universal particle $\mathscr{H}$ are equivalent:

(a) $\mathscr{H}$ is elementary;

(b) $\mathscr{H}$ does not have any closed linear subspace G-invariant except $\{0\}$ and $\mathscr{H}$ itself;

(c) If a G-invariant Hilbert space $\mathscr{H}_1$, continuously embedded in $\mathscr{D}'(V) \otimes \mathbf{F}$, is contained in $\mathscr{H}$, then $\mathscr{H}_1 = \mathscr{H}$ with a proportional norm.

Condition (b) can be rephrased as follows: the representation of G in $\mathscr{H}$ (that is, in the group of unitary operators of $\mathscr{H}$) is irreducible.

We have trivially $(a) \to (b)$ and $(c) \to (a)$. We shall prove $(b) \to (a)$ and $(a) \to (c)$.

$(b) \to (a)$. Let $\mathscr{H}_1 \leqslant \mathscr{H}$, $\mathscr{H}_1 \neq \{0\}$, be G-invariant. To $\mathscr{H}_1$ corresponds (in a unique manner) an anti-kernel L_1: $\mathscr{H}' \to \mathscr{H}$. Let L

be the canonical anti-isomorphism $\mathscr{H}' \to \mathscr{H}$. For $x \in \mathscr{H}$, let $x' = L^{-1}x$. For $y \in \mathscr{H}_1$ we have:

$$(y|x)_{\mathscr{H}} = (y|Lx')_{\mathscr{H}} = \langle y, x' \rangle = (y|L_1 x')_{\mathscr{H}_1}.$$

Let us set $Ax = L_1 L^{-1}x$. One has:

$$(y|x)_{\mathscr{H}} = (y|Ax)_{\mathscr{H}_1}, x \in \mathscr{H}, y \in \mathscr{H}_1.$$

Since any $\sigma \in G$ operates unitarily on both $\mathscr{H}$ and $\mathscr{H}_1$

$$(y|\sigma Ax)_{\mathscr{H}_1} = (\sigma^{-1} y|Ax)_{\mathscr{H}_1} = (\sigma^{-1} y|x)_{\mathscr{H}} = (y|\sigma x)_{\mathscr{H}} = (y|A\sigma x)_{\mathscr{H}_1}.$$

Thus A commutes with G. Therefore every spectral manifold of A is invariant by G (note that A is a bounded operator $\mathscr{H} \to \mathscr{H}$ and $A \geqslant 0$, for

$$(Ax|x)_{\mathscr{H}} = [(Ax|Ax)_{\mathscr{H}_1} \geqslant 0].$$

But according to (b) there are no non-trivial closed subspaces of $\mathscr{H}$ G-invariant. Hence A must have only one spectral manifold, $\mathscr{H}$ itself. For some $\lambda \in C$, $Ax = \lambda x$ whatever be $x \in \mathscr{H}$ and:

$$(y|x)_{\mathscr{H}} = \bar{\lambda}(y|x)_{\mathscr{H}_1}$$

Take $x = y \in \mathscr{H}_1$; first, $\bar{\lambda} = (y|y)_{\mathscr{H}_1}/(y|y)_{\mathscr{H}}$ must be a number > 0; and:

$$\| y \|_{\mathscr{H}_1} = \frac{1}{\sqrt{\lambda}} \| y \|_{\mathscr{H}}$$

(a) $\to$ *(c)*. Suppose that $\mathscr{H}_1 \subset \mathscr{H}$ is another G-invariant Hilbert space with continuous injection in $\mathscr{D}'(V) \otimes \mathbf{F}$. Then the natural embedding of $\mathscr{H}_1$ into $\mathscr{H}$ must be continuous, according to the closed graph theorem. This means that there is a constant $M < +\infty$ such that, for every $x \in \mathscr{H}_1$,

$$\| x \|_{\mathscr{H}} \leqslant M \| x \|_{\mathscr{H}_1}.$$

Let us define a new Hilbert space $\mathscr{H}_2$ as follows: as a set, $\mathscr{H}_2 = \mathscr{H}_1$; and for $x \in \mathscr{H}_1$,

$$\| x \|_{\mathscr{H}_2} = M \| x \|_{\mathscr{H}_1}$$

Now $\mathscr{H}_2 \leqslant \mathscr{H}$: $\mathscr{H}_2 \subset \mathscr{H}$ and $\| x \|_{\mathscr{H}} \leqslant \| x \|_{\mathscr{H}_2}$, $x \in \mathscr{H}_2$. According to (a), $\mathscr{H}_2 = \mathscr{H}$ as a set and there is a constant $\lambda \geqslant 0$ such that, for every $x \in \mathscr{H}$,

$$\|x\|_{\mathscr{H}_2} = \lambda \|x\|_{\mathscr{H}}.$$

Hence, $\mathscr{H}_1 = \mathscr{H}$ and $\|x\|_{\mathscr{H}_1} = \dfrac{\lambda}{M}\|x\|_{\mathscr{H}}$. Q.E.D.

Let us go back to the particular case $V = E_n$ and G satisfying the conditions of the beginning. The fact that the particle $\mathscr{H}$ is elementary can be translated as follows:

(1) *To the anti-kernel K corresponding to $\mathscr{H}$*. If $K_1 \leqslant K$ is another positive anti-kernel G-invariant, then $K_1 = \lambda K$ for some $0 \leqslant \lambda \leqslant 1$;

(2) *To the distribution $H_{\mathbf{u}}$*. If H_1 is another B-positive (or B-superpositive) distribution G-invariant and $H_1 \leqslant H$ in the sense of Bochner-positivity, then $H_1 = \lambda H$;

(3) *To the measure μ*. If μ_1 is another tempered positive measure G-invariant and $\mu_1 \leqslant \mu$ in the sense of measures, then $\mu_1 = \lambda\mu$.

The collection $\mathfrak{H}$ of the $\mathbf{F}$-particles form a closed convex cone. Those G-invariant form a subcone. The elementary particles form the extremal generatrices of this subcone.

Supports of Extremal Measures

Consider the case in which the manifold $V = E_n$, an n-dimensional affine space over the reals (F, as usual, is a finite dimensional vector space over the complex field). We have a group G operating on E_n as an affine group and on F as a linear group, having the properties:

(1) All the elements of G preserve Lebesgue measure on E_n, that is, each operator in G has det $= \pm 1$ as an operator on $\mathbf{E}_n$.

(2) G covers all the translations, that is, for any translation $\mathbf{a} \in \mathbf{E}_n$ there exists $g_{\mathbf{a}} \in G$ which operates as the translation $\mathbf{a}$ on E_n and identically on F.

Let G_0 be the subset of G of all the elements that operate as translations on E_n and identically on F; G_0 forms an invariant subgroup.

We are looking for a Hilbert space $\mathscr{H} \subset \mathscr{D}'(E_n) \otimes F$ with a finer topology than the induced one and G-invariant. This is equivalent to looking for a tempered measure μ on E_n' having values in $F \otimes \overline{F}$, positive, and G-invariant. If we want *elementary particles*, we also need that $\mathscr{H}$ be extremal, that is, if we have any other space $\mathscr{H}_1 \subset \mathscr{H}$ analogous to $\mathscr{H}$ with a greater norm (or simply just $\mathscr{H}_1 \subset \mathscr{H}$) and G-invariant, then $\mathscr{H}_1 = \mathscr{H}$ with a proportional norm, or, equivalently, if we have $\mathscr{H}_1 \leqslant \mathscr{H}$ and G-invariant, then $\mathscr{H} = \mathscr{H}$ with a proportional norm. Finally, the condition on μ is that it is extremal, that is, if μ_1 is another tempered positive measure, G-invariant, such that $\mu_1 \leqslant \mu$, then μ_1 is proportional to μ.

We shall give now a necessary condition for a measure to be extremal.

Definition. Let G be a group operating on a topological space. A *G-orbit* is the set of all transforms by G of one point of the space.

Theorem. Let μ be a measure on a locally compact space X with a countable basis of open sets, and let the values of μ lie in a finite dimensional ordered vector space $\mathscr{F}$. If μ is positive, G-invariant (or G/G_0 invariant), and extremal among the measures having these properties, then the support of μ is the closure of one G-orbit.

Remarks. For the case in which we are interested, $X = \mathbf{E}_4'$, $\mathscr{F} = F \otimes \overline{F}$ and G/G_0 acts in E_4 as the homogenous Lorentz group. An orbit is, for instance, a hyperboloid, since the Lorentz group preserves the quadratic form

$$p^2 = p_1^2 + p_2^2 + p_3^2 - p_0^2.$$

Then we are sure that such a measure necessarily has its support in one hyperboloid. This gives us a considerable restriction on the supports of the measures which may be extremal and it will not be difficult to find these measures. In fact, we shall show that in the case of an hyperboloid there is only one measure, up to a constant factor.

Note that we have no choice but to take the *closure* of the orbit, for a support is always closed. There do exist G-orbits which are not closed. For example, the upper part of the surface of the light cone without its vertex (the origin) is such a G-orbit. The closure of this orbit is the upper cone with the vertex included, and this may be the support of a measure. However, we can still have the measure *concentrated* on the cone without its vertex in the sense of the following definition.

Definition. A measure is concentrated on a subset if the measure of the complement of this set has measure zero.

In our example, a measure whose support is the cone *with its vertex* is concentrated in the cone *without the vertex* whenever the measure of the vertex is zero.

It would be a stronger statement to say that the measure is *concentrated* on one G-orbit. But this is not possible. First of all, the notion of concentration has no uniqueness character. For example, the Lebesgue measure on the straight line is cóncentrated on the complement of *any* countable subset. In addition, the theorem would be false if we insisted that the measure should be concentrated on one orbit, as shown by the following example: Let X be a circle, and G be the group of rotations generated by one rotation incommensurable with π. Any orbit is a dense, countable set, and the only invariant measure under G is the Lebesgue measure on the circle, because if a measure is invariant over a group it is also invariant over the closure of the group, which in this case is the group of all rotations. Thus, there is no measure concentrated in the G-orbit in this case.

Proof of theorem. It is sufficient to let X = support of μ, because if μ is invariant, the support of μ is also invariant. We must prove that there exists one orbit dense in X. Take any point $a \in X$, and let V be any open neighborhood of a. Take the saturation $\tilde{V}$ of V, that is, the union of the transforms of V by all elements of G. The set $\tilde{V}$ is an open set and has a positive measure because any non-empty open set has a positive measure (otherwise support $\mu \neq X$). The complement of $\tilde{V}$ has measure zero; otherwise, we could multiply the measure μ by the characteristic function of $\tilde{V}$ and obtain a positive G-invariant measure $\leqslant \mu$, not proportional to μ. Thus, $\tilde{V}$ is almost all X. The point a has a countable fundamental system of neighborhoods V_n. Each set $\tilde{V}_n$ is almost all of X. Then $\cap \tilde{V}_n$ is still almost all of X. In other words, $b \in \cap \tilde{V}_n$ for almost all $b \in X$.

If $b \in \cap \tilde{V}_n$, then $b \in \tilde{V}_n$ for all n; thus the orbit $\tilde{b}$ of b meets all $\tilde{V}_n$; that is, $\tilde{b}$ meets all the neighborhoods of a. Therefore, $a \in \bar{\tilde{b}}$ (closure of b) for almost all b. Take a countable set $\{a_k\}$ dense in X. For each fixed k, $a_k \in \bar{\tilde{b}}$ for almost all b. Since $\{a_k\}$ is countable, it follows that $a_k \in \bar{\tilde{b}}$ for every k for almost all $\tilde{b}$. Therefore $\bar{\tilde{b}} = X$ for almost all $b \in X$.

Mesons

We are going to apply now the previous results to the theory of the meson. In this case $V = E_4$ and $F = C$; G operates identically on F and as the *proper inhomogeneous Lorentz group* on E_4.

We recall that the inhomogeneous Lorentz group has four connected components. This is true of the *homogeneous* Lorentz group $\mathscr{L}$:

(1) Transformations which preserve the sense of time and with determinant $+1$;

(2) Transformations which preserve the sense of time but have determinant -1;

(1′) Transformations which invert the sense of time and with determinant $+1$;

(2′) Transformations which invert the sense of time and with determinant -1.

Since $\mathscr{L}$ is the quotient of the inhomogeneous Lorentz group by the (abelian invariant) subgroup of translations, the connected components of the inhomogeneous Lorentz group are the preimages of the connected components of $\mathscr{L}$.

Here we take for G the pre-image of the set (1) (which is called *the proper homogeneous Lorentz group*). Observe that the set (1) is the only connected component of $\mathscr{L}$ which is a subgroup; it is the connected component of the identity.

The meaning of preservation or inversion of the sense of time can be explained with the light cone.

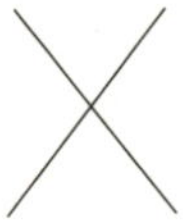

The light cone will be mapped onto itself by any $\sigma \in \mathscr{L}$. There are two possibilities:

(1) σ maps the upper part of the light cone onto itself and the lower part onto itself. In this case, the sense of time is said to be preserved by σ.

(2) σ maps the upper part of the light cone onto the lower part and the lower part onto the upper part. In this case, the sense of time is said to be inverted by σ.

One may tell if the sense of time is preserved by taking a coordinate system and representing σ by a matrix. Then the transformed time component is

$$X'_0 = C_{01}X_1 + C_{02}X_2 + C_{03}X_3 + C_{00}X_0$$

and C_{00} cannot be zero. If $C_{00} > 0$, σ preserves the sense of time, and if $C_{00} < 0$, σ inverts the sense of time.

The choice of the group G is of fundamental importance since it determines what the elementary particles will be. Mathematically, many choices are possible, but physically, there is a reason for choosing the proper inhomogeneous Lorentz group. First of all, it is physically reasonable to have a sense of orientation and a sense of time, at least locally. That makes it natural to take the proper inhomogeneous Lorentz group instead of the entire inhomogeneous Lorentz group. For example, if we had the whole Lorentz group [or just the part which is the pre-image of the transformations (1) and (2′)], we would have charged particles but with no well-determined charge; that is, we would have a probability of having one charge or the other.

However, this is not a sufficient reason for choosing the proper inhomogeneous Lorentz group. Observe that this group is not the one used by the Special Theory of Relativity, where no fundamental unit of length is given. We do not have E_4 with a given quadratic form, but instead we have E_4 with a family of proportional quadratic forms. Thus it seems that the correct group should include all the dilatations (in addition to the proper Lorentz group). If this were the correct group, the elementary particles would be completely different. There will be types of particles with no determined mass, but with all the masses given with a law of probability (probability that a measurement would find a certain mass).

It appears that microphysics gives us a reason for excluding the dilatations. Physical particles, like the electron or proton, for example, actually do not have all possible masses, and not all wave lengths are found in the atomic spectra of each element. As soon as we decide that there are privileged masses or lengths in the universe, then the group we must choose is the proper inhomogeneous Lorentz group.

Let us set G_0 equal to a group of translations. The factor group G/G_0 will operate on $\mathbf{E}'_4$, and all orbits will be contained in hyperboloids:

$$p^2 = a, \quad a = \text{constant},$$

since p^2 is invariant under G/G_0.

There are three kinds of hyperboloids:

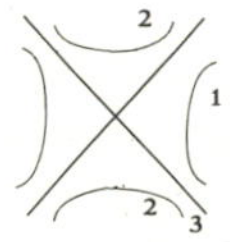

(1) having one connected component
(2) having two connected components
(3) the light cone

A G/G_0 orbit is contained in one of these hyperboloids but is not necessarily the whole of it.

However, any orbit contained in an hyperboloid of type (1), that is, a one-branch (or connected) hyperboloid, necessarily coincides with the hyperboloid itself. For, take any point $\mathbf{a}$ of this hyperboloid; by choosing a suitable system of coordinates, we may assume that this point is $(a_1, 0,0,a_0)$. Given any other point $\mathbf{b} = (b_1,b_2,b_3,b_0)$ there is a space rotation ρ that maps it on a point of the type $(b_1', 0,0,b_0)$. But then there is a real number θ such that

$$\begin{aligned} a_1 &= b_1' \cos h\theta + b_0 \sin h\theta, \\ a_0 &= b_1' \sin h\theta + b_0 \cos h\theta \end{aligned}$$

Let us denote by σ_θ this transformation; we see that $\sigma_\theta \rho$ maps $\mathbf{b}$ onto $\mathbf{a}$, and $\sigma_\theta \rho \in G/G_0$ quite obviously.

It can be shown in the same way that there are two kinds of orbits in (2): the upper sheet of (2) and the lower sheet of (2). If, on the other hand, G included also transformations not preserving the sense of time (for example, if G were the whole Lorentz group), then there would be only one orbit in (2), because both sheets together would be one orbit. For the light cone (3), there are three kinds of orbits: the upper part without the origin, the lower part without the origin, and the origin.

First, we shall examine the Hilbert space corresponding to the simplest kind of orbit, the origin. All the positive tempered measures located at the origin and extremal are proportional to δ:

$$\mu = c\delta, \qquad c > 0$$

Proportional Hilbert spaces will define the same particle, that is, a particle is a class of proportional Hilbert spaces or a whole generatrix of the closed convex cone. Thus, we may take $\mu = \delta$ because all the other measures are proportional. $H = 1$ is the inverse Fourier transform of μ. The space L_μ^2 is the space of all constants. The space $\Lambda_\mu^2 = \mathscr{F}\mathscr{H}$ is the space of all $f\delta$ with f an arbitrary constant (that is, $f \in L_\mu^2$) and with the inner product $(f\delta | g\delta)_{\mathscr{F}\mathscr{H}} = f\bar{g}$. The inverse Fourier transform of this space is $\mathscr{H}$, which is the space of constant functions. The inner product of α, $\beta \in \mathscr{H}$ is $\alpha\bar{\beta}$. The physical interpretation of this space $\mathscr{H}$ is not quite clear. We shall call it the *vacuum* of mesons.

Lorentz Invariant Distributions

In looking for positive tempered measures having their support on a given orbit and extremal, we shall make use of some results of P. D. Méthée. Méthée has given a general expression for the distributions which are invariant under the homogenous extended Lorentz group $\mathscr{L}$.

First Theorem of Méthée.

The mapping $f(u) \to f(p^2)$ of continuous functions on R^1 into continuous functions on E_4, invariant under $\mathscr{L}$, can be extended in a unique manner into a linear topological isomorphism $T_u \to T_{p^2}$ of $\mathscr{D}'(R^1)$ into $\mathscr{D}'(\mathbf{E}_4' - \{0\})$. The image of $\mathscr{D}'(R^1)$ under this map consists of distributions on $\mathbf{E}_4' - \{0\}$ invariant under $\mathscr{L}$.

Remarks. The proof of the existence of the map will not be given here. The uniqueness follows from the fact that continuous functions on R^1 are dense in $\mathscr{D}'(R^1)$.

The theorem gives a notion of an inverse image of a distribution for the map $\mathbf{p} \to p^2$ from $\mathbf{E}_4' - \{0\}$ into R^1: T_{p^2} is the inverse image of the distribution T_u. The support of any T_{p^2} is easily found: it is the inverse image of the support of T by the map $\mathbf{p} \to p^2$.

Example. δ_{p^2} exists in $\mathbf{E}_4' - 0$. The support of δ_{p^2} is the surface of a light cone excluding the origin.

This theorem is a particular case of a more general theorem which states that if an infinitely differentiable map from one manifold, V, into another, W, is of constant rank, then one can always take the inverse image of a distribution on W; the inverse image is a distribution on V.

Second Theorem of Méthée.

Any distribution in $\mathbf{E}_4' - \{0\}$ invariant under $\mathscr{L}$ is of the form T_{p^2}, with $T_u \in \mathscr{D}'(R^1)$ (i.e., the isomorphism in the first theorem maps $\mathscr{D}'(R^1)$ *onto* the space of invariant distributions on $\mathbf{E}_4' - \{0\}$).

Remarks. Now if we consider all of $\mathbf{E}_4'$ instead of $\mathbf{E}_4' - \{0\}$, this theorem does not apply. If we have a distribution T_u, this theorem does not say that there necessarily corresponds a T_{p^2} in the whole space $\mathbf{E}_4'$. Moreover, if this correspondence does exist, we cannot get in any well defined way all the Lorentz invariant distributions over $\mathbf{E}_4'$. For any Lorentz invariant distribution with its support at the origin cannot be obtained in this way. That is why we consider only the complementary set of the origin, $\mathbf{E}_4' - \{0\}$.

Now we want to find distributions in $\mathbf{E}_4' - \{0\}$ which are invariant under the proper homogenous Lorentz group $\mathscr{L}_+^\uparrow$ instead of the extended homogenous Lorentz group $\mathscr{L}$. First we must state a special case of Méthée's first theorem and a stronger form of Méthée's second theorem. Let Ω_+ be the open set obtained by subtracting the lower part of the light cone and the origin from $\mathbf{E}_4'$

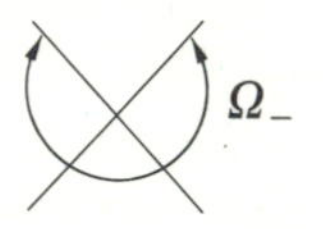

Theorem. The mapping $f(u) \to f(p^2)$ of continuous functions on R^1 into restrictions to Ω_+ of continuous functions on E_4 (invariant under $\mathscr{L}$) can be extended in a unique way into a linear topological isomorphism $T_u \to T_{p^2}$ of $\mathscr{D}'(R^1)$ into $\mathscr{D}'(\Omega_+)$. The image of $\mathscr{D}'(R^1)$ under this map consists of *all* distributions on Ω_+ invariant under the proper homogeneous Lorentz group.

Remarks. We have stated together, in this special case, what had been divided into two parts (First and Second Theorem of Méthée) in the case of $\mathbf{E}_4' - \{0\}$. The proof will not be given here. The corresponding theorems in which Ω_+ is replaced by Ω_- is also true, where Ω_- is the open set obtained by subtracting the upper part of the light cone and the origin from $\mathbf{E}_4'$.

Note that Ω_+ and Ω_- are invariant under the proper homogeneous Lorentz group.

Theorem. Let T be any distribution on $\mathbf{E}_4' - 0$ which is invariant under the proper homogeneous Lorentz group $\mathscr{L}_+^\uparrow$. Then in Ω_+, T is the inverse image $T_{p^2}^+$ of a distribution T_u^+ on R^1 and in Ω_- it is the inverse image $T_{p^2}^-$ of a distribution T_u^- on R^1, where T_u^+ and T_u^- coincide in the region $u > 0$ in R^1.

Converse. Given two distributions T_u^+, T_u^- on the straight line R^1 which coincide on $u > 0$, their inverse images $T_{p^2}^+$, $T_{p^2}^-$ coincide on $\Omega_+ \cap \Omega_-$ and the pair $(T_{p^2}^+, T_{p^2}^-)$ defines a distribution on $\mathbf{E}_4' - \{0\}$ which is invariant under the proper homogeneous Lorentz group $\mathscr{L}_+^\uparrow$.

Proof of Theorem. Let T be any distribution on $\mathbf{E}_4' - \{0\}$ which is invariant under $\mathscr{L}_+^\uparrow$. On Ω_+, T defines a distribution which is invariant under $\mathscr{L}_+^\uparrow$. By the special case of the Second Theorem of Méthée, T on Ω_+ is the inverse image $T_{p^2}^+$ of a distribution T_u^+ on R^1. Similarly, T defines a distribution on Ω_- which is invariant under $\mathscr{L}_+^\uparrow$ and T

on Ω_- is the inverse image $T^-_{p^2}$ of a distribution T^-_u in R. In the region $\Omega_+ \cap \Omega_-$, $T^+_{p^2}$ and $T^-_{p^2}$ must coincide. Since $\Omega_+ \cap \Omega_-$ is the inverse image of the region $u > 0$ in R^1. T^+_u and T_u must coincide for $u > 0$ in R.

Proof of the converse. Take two arbitrary distributions T^+_u and T^-_u which coincide for $u > 0$ in R^1. By the First Theorem of Méthée (special case), T^+_u defines a distribution $T^+_{p^2}$ on Ω_+ which is invariant under $\mathscr{L}^{\uparrow}_{+}$. Likewise, T^-_u defines a distribution $T^-_{p^2}$ on Ω_- which is invariant under $\mathscr{L}^{\uparrow}_{+}$. Since T^+_u and T^-_u coincide for $u > 0$ in R, their inverse images $T^+_{p^2}$ and $T^-_{p^2}$ coincide on $\Omega_+ \cap \Omega_-$, inverse image of the set $u > 0$. Thus $(T^+_{p^2}, T^-_{p^2})$ define a distribution on $\mathbf{E}'_4 - \{0\} = \Omega_+ \cup \Omega_-$ which is invariant under $\mathscr{L}^{\uparrow}_{+}$.

We see then that to find all the distributions on $\mathbf{E}'_4 - \{0\}$ which are invariant under the proper homogeneous Lorentz group, it is equivalent to find all pairs of distributions on R^1 which coincide for $u > 0$.

Determination of all Mesons

Consider a hyperboloid of one sheet (1).

(1)

It has the equation $p^2 = k^2$, where k^2 is a positive constant. We are looking for a measure having its support on (1) and which is invariant under the proper homogeneous Lorentz group. By the results following from Méthée's theorems it must be defined by two distributions $T^+_{p^2}$, $T^-_{p^2}$ which are inverse images of two distributions T^+_u, T^-_u which coincide on $u > 0$ in R. According to the law of supports, the supports of T^+_u and T^-_u must be the point k^2 on R. We have the two measures T^+_u and T^-_u having their supports at k^2, and which must coincide for $u > 0$, thus they must coincide on the entire real line. In Méthée's theorems, the order of the distribution is preserved by the isomorphisms; thus to obtain a measure $\geqslant 0$ on $\mathbf{E}'_4 - \{0\}$ we must take a measure $\geqslant 0$ on R^1. Now the only positive measure with its support at k^2 is δ_{u-k^2} to a constant factor. Finally, the measure on $\mathbf{E}'_4 - \{0\}$ is defined by the two Méthée distributions $\delta_{p^2-k^2}$, $\delta_{p^2-k^2}$, and since they are the same, only one distribution, $\delta_{p^2-k^2}$, is needed to define our positive measure in $\mathbf{E}'_4 - \{0\}$. Since the support is the hyperboloid (1), which does not contain the origin, it follows that there is only one invariant distribution in $\mathbf{E}'_4$ that has (1) for its support. This distribution is defined by the two distributions $\delta_{p^2-k^2}$ on $\mathbf{E}'_4 - \{0\}$ and 0 on $\mathbf{E}'_4 - \{\mathbf{p}\colon p^2 = k^2\}$. These two distributions are defined on open sets, they coincide on the intersection of the open sets, and $\mathbf{E}'_4$ is the union of these open sets. Thus on $\mathbf{E}'_4$ we have defined a distribution which we may call $\delta_{p^2-k^2}$.

From a theorem proved by Méthée it follows that the isomorphism $T_u \longleftrightarrow T_{p^2}$ preserves also the tempered character. Thus a Lorentz invariant or proper Lorentz invariant distribution is tempered if and only if the corresponding distribution (or pair of distributions) on the straight line is (or are) tempered. Therefore $\delta_{p^2-k^2}$ is tempered since δ_{u-k^2} is tempered. We now know that with this hyperboloid (1) as support, there exists one and only one positive measure in $\mathbf{E}'_4$ (i.e., $\delta_{p^2-k^2}$) to a constant factor which is invariant under the proper homogeneous Lorentz group. This measure is extremal because only one exists to a constant factor. If we take the corresponding class of Hilbert spaces we see that each hyperboloid of one sheet defines one and only one particle.

These particles are not acceptable physically if the hypothesis of positive energy is assumed.

Now consider the hyperboloid of two sheets.

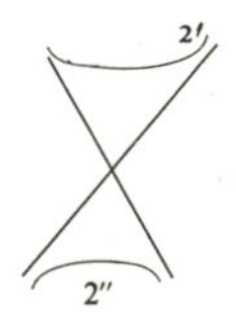

There are two orbits $2'$ and $2''$. By the same reasoning as before we find that any distribution which is proper Lorentz invariant and has the support $2'$ is defined by the two distributions

$$\delta_{p^2+k^2}, 0$$

in the sense of Méthée. Similarly any distribution which is proper Lorentz invariant and has the support $2''$ is given by

$$0, \delta_{p^2+k^2}.$$

By the preservation of tempered character under the correspondence given by Méthée, these distributions are tempered. Thus each sheet of a hyperboloid (2) of two sheets gives a particle, i.e., we have particles depending on a parameter $k^2 > 0$ and a sign $\pm$, conventionally $-$ if we take the upper sheet, and $+$ if we take the lower sheet.

Such particles are studied in physics and physicists set

$$k^2 = \frac{m_0^2 c^2}{h^2},$$

where m_0 is called the *rest mass* of the particle, c is the *velocity of light*, and h is *Planck's constant.*

For mass 0, i.e., $k^2 = 0$, the problem is a little more difficult. The upper part of the surface of the light cone without the origin is one orbit. Likewise, the lower part of the surface of the light cone without the origin is another orbit. The measure

μ in $\mathbf{E}_4' - 0$ with support in the upper part of the cone is given by the Méthée distributions

$$\delta_{p^2}, 0.$$

Now we want to extend this measure to $\mathbf{E}_4'$. The support of this extended measure, if it exists, will be the surface of the upper part of the light cone with the origin included. The extension of this measure to $\mathbf{E}_4'$ will exist only if the measure over any compact set including the origin will have a finite mass. But this measure can be finite if and only if the dimension n of the space is $\geqslant 3$. Since $n = 4$ in this case, the measure can be extended to $\mathbf{E}_4'$. However, this can be done in an infinite number of ways, for any constant times a Dirac measure δ at the origin can be added. Note that if $n = 2$, then there exists no particle of mass zero.

The measure μ can be extended in a unique way to $\mathbf{E}_4'$ if we want the origin $\{0\}$ to have measure zero. The extension is given by

$$\mu(\varphi) = \int \varphi \, \mathrm{d}\mu,$$

where the integral is taken over $\mathbf{E}_4' - \{0\}$; φ is any continuous function on $\mathbf{E}_4'$ with compact support. Now an arbitrary constant times a Dirac measure may be added to this measure. But if we impose the condition that the measure be extremal, we see that the origin must have measure zero. For $\lambda\delta \leqslant \mu + \lambda\delta$.

Similarly, the measure with support in the lower part of the cone can be extended to $\mathbf{E}'_4$.

To summarize, we see that there are three kinds of particles:

(1) One special particle, the *vacuum*;
(2) Particles defined by $p_2 = k^2, k^2 > 0$;
(3) Particles defined by $p^2 = -k^2, \pm, k^2 \geqslant 0$.

The particles (3) are physically acceptable. Note, however, that no positive masses are rejected. Up to now, only mesons with masses belonging to a discrete set of positive numbers are known. Hence, either an additional principle of selection of masses is needed, or we should expect more and more mesons of arbitrary masses to be discovered.

Description of $\mathscr{H}$ for the Meson

Now the space $\mathscr{H}_{m_0}, \pm$ will be described in more detail. This Hilbert space represents a particle with rest mass m_0 and charge plus or minus. We have considered hyperboloids in $\mathbf{E}'_4$ defined by the equation

$$p^2 + m^2 = 0$$

where

$$m = \frac{m_0 c}{h}.$$

The upper and lower sheets correspond to minus and plus charges, respectively. On each branch of this hyperboloid there exists one and only one positive Lorentz invariant measure to a constant factor. For example, on the plus branch the measure is given in the sense of Méthée by the inverse image of the pair of distributions

$$(0, \delta_{u+m^2})$$

on R^1, the inverse image of which is

$$(0, \delta_{p^2+m^2}).$$

This measure is also classically written

$$\mu = Y(-p_0)\,\delta(p^2 + m^2)$$

where Y is the Heaviside function. The "Fourier transform" $\mathscr{F}\mathscr{H}$ of

$\mathscr{H}$ has a meaning if we choose an origin in E_4; $\mathscr{F}\mathscr{H}$ is the Hilbert space Λ_μ^2 of distributions on E_4 of the form $f\mu$, where f is square-integrable with respect to the measure μ.

The scalar product in Λ_μ^2 is

$$(f\mu|g\mu)_{\mathscr{F}\mathscr{H}} = \int f\bar{g}\,\mathrm{d}\mu.$$

The distribution $H \in \mathscr{S}'(\mathbf{E}_4)$ is given by $H = \overline{\mathscr{F}}\mu$ and the corresponding anti-kernel K is defined by $K_{x,\xi} = H_{\overrightarrow{x-\xi}}$. Méthée has computed some inverse Fourier transforms, which we may use for finding an expression for H. Let us introduce the distribution:

$$2\pi\Delta'_{2\pi m_0 c/h}(x) = \overline{\mathscr{F}}\delta\left(p^2 + \frac{m_0^2c^2}{h^2}\right)$$

$$= pv\left\{\frac{m_0 c\pi}{h}\frac{N_1[2\pi(m_0c/h)\sqrt{-x^2}]}{\sqrt{-x^2}}Y(-x^2)\right.$$

$$\left. + \frac{2m_0c}{h}\frac{K_1[2\pi(m_0c/h)\sqrt{x^2}]}{\sqrt{x^2}}Y(x^2)\right\},$$

where pv denotes Cauchy's principal value. By definition of pvf, with f denoting the function above, if φ is a testing function, one has:

$$\langle pvf, \varphi\rangle = \lim_{\varepsilon\to 0}\int_{|x^2|>\varepsilon} f\varphi\,\mathrm{d}x$$

N_1 is Neumann's function, and K_1 is Kelvin's function.

Also:

$$2\pi i\Delta_{2\pi m_0c/h}(x) = \overline{\mathscr{F}}\left[\varepsilon(p_0)\,\delta\left(p^2 + \frac{m_0^2c^2}{h^2}\right)\right]$$

$$= i\varepsilon(x_0)\left\{-\frac{m_0c\pi}{h}\frac{J_1[2\pi(m_0c/h)\sqrt{-x^2}]}{\sqrt{-x^2}}Y(-x^2) + \delta(x^2)\right\}$$

where

$$\varepsilon(p_0) = \begin{cases} 1 \text{ for } p_0 > 0 \\ -1 \text{ for } p_0 < 0\end{cases}$$

and J_1 is the Bessel function. One may then prove:

$$H_{m_0,+} = 2\pi\Delta^+_{2\pi m_0c/h}(x)$$

where

$$\Delta^{+} = \frac{\Delta' - i\Delta}{2}.$$

Fourier transforms preserve even and odd symmetry with respect to the origin and change bar (complex conjugation) into tilda ($\tilde{\varphi}(x) = \overline{\varphi(-x)}$). Now

$$\delta\left(p^2 + \frac{m_0^2c^2}{h^2}\right)$$

is even, and since it is also real, it is invariant by tilda. Therefore Δ' is even and invariant by bar, i.e., real. Similarly, it follows that $i\Delta$ is odd and pure imaginary.

All the $\psi \in \mathscr{H}$ will be functions, but H on the other hand is a distribution. All $\psi \in \mathscr{H}$ satisfy the Klein–Gordon equation:

$$\Box\ \psi - \frac{4\pi^2 m^2 c^2}{h^2}\psi = 0.$$

This is proved by taking the Fourier transform $\mathscr{F}\psi = f\mu$, which is a measure carried by the hyperbloid

$$p^2 + \frac{m_0^2c^2}{h^2} = 0.$$

This measure is annihilated by multiplication with

$$p^2 + \frac{m_0^2c^2}{h^2},$$

$$p^2 + \frac{m_0^2c^2}{h^2}\ (f\mu) = 0.$$

The inverse Fourier transform of this equation is the Klein–Gordon equation. Because of the obtention by means of Fourier transform, the d'Alembertian has to be taken in the sense of distributions; ψ does not, in general, have second derivatives in the usual sense.

$\mathscr{H}$ does not contain all the solutions of the Klein–Gordon equation. There exist solutions which are not tempered, therefore not in $\mathscr{H}$. Now consider an arbitrary tempered solution of the Klein–Gordon equation. Its Fourier transform has its support in the union of the two branches of the hyperboloid of two sheets. Only in special cases will the support be only one branch; thus not all tempered solutions of the Klein–Gordon equation are in $\mathscr{H}$. But there is still the additional restriction that the Fourier transform of elements of $\mathscr{H}$ be a measure of the form $f\mu$ where f is μ square-integrable.

CHAPTER 5

Spin Particles

Vector Particles

Let $V = E_4$, $\mathbf{F}$ = finite-dimensional vector space over C. Let the structure group G satisfy the following hypotheses:

(1) G operates on E_4 as the inhomogeneous proper Lorentz group, i.e., a representation of G into the affine group of E_4 is given, whose image is the inhomogeneous proper Lorentz group.

(2) For any translation $\mathbf{a} \in \mathbf{E}_4$, there exists $g \in G_{\mathbf{a}}$ which operates as this translation $\mathbf{a}$ on E_4 and operates identically on $\mathbf{F}$.

Note that G preserves Lebesgue measure because of hypothesis (1).

Consider an elementary particle $\mathscr{H} \subset \mathscr{D}'(V; \mathbf{F})$. By G operating on $\mathbf{F}$, we mean that we have in $\mathbf{F}$ a linear representation of G. One usually says that a distribution $T \in \mathscr{D}'(V; \mathbf{F})$ takes its values in a subspace $\mathbf{F}_1$ of $\mathbf{F}$ if for every $\varphi \in \mathscr{D}(V)$, $\langle \mathbf{T}, \varphi \rangle \in \mathbf{F}_1$.

Theorem. Let $\mathscr{H}$ be an elementary F-particle. We make the two following assumptions:

(1) There is no subspace F_1 of F, $F_1 \neq F$, such that all $\psi \in \mathscr{H}$ take their values in F_1.

(2) The representation of G in F is reducible. Then there exists a linear subspace F_1 of F, G-invariant, with the following properties:

(I) There is an isometry of $\mathscr{H}$ onto an F^0-particle $\mathscr{H}^0$, with $F^0 = F/F_1$.

(II) The quotient representation modulo F_1 of G in F^0 is irreducible.

Proof. Let F_1 be a proper subspace of F, G invariant. Let $\mathscr{H}_1$ be the subspace of all $\psi \in \mathscr{H}$ taking their values in F_1; $\mathscr{H}_1$ is a closed linear subspace of $\mathscr{H}$. Since F_1 is G-invariant, $\mathscr{H}_1$ is also G-invariant. Since the particle $\mathscr{H}$ is elementary, we should have either $\mathscr{H}_1 = \mathscr{H}$ or $\mathscr{H}_1 = \{0\}$. But we cannot have $\mathscr{H}_1 = \mathscr{H}$ according to (1). Hence $\mathscr{H}_1 = \{0\}$. Let π be the canonical projection of F onto $F^0 = F/F_1$. Every ψ with values in F defines canonically a distribution $\dot{\psi}$ with values in F^0 by means of the equation

$$\langle \dot{\psi}, \varphi \rangle = \pi(\langle \psi, \varphi \rangle), \varphi \in \mathscr{D}(V).$$

For $\sigma \in G, \dot{f} \in F^0$, we define

$$\sigma \dot{f} = (\sigma f)^0,$$

which is independent of the representative $f \in \dot{f}$ because F_1 is invariant under σ. Let $\mathscr{H}^0$ be the set of all $\dot{\psi}$ when ψ runs over $\mathscr{H}$. Of course $\mathscr{H}^0 \subset \mathscr{D}'(V; F^0)$ and the group G operates in $\mathscr{H}^0$. If $\dot{\psi} = 0$ then ψ takes its values in F_1, and therefore $\psi = 0$, since $\mathscr{H}_1 = \{0\}$. Hence the mapping $\psi \to \dot{\psi}$ from $\mathscr{H}$ onto $\mathscr{H}^0$ is one to one. We may transfer the structure of $\mathscr{H}$ on $\mathscr{H}^0$. If the representation of G in F^0 is reducible, then we may continue in the same way and apply the previous reasoning to F^0 instead of F. Since $\dim F < +\infty$, this process will lead us in a finite number of steps to an F^0 with property (II).

This theorem shows that we may restrict ourselves to elementary F-particles with the assumption that the representation of the structural group G in F is irreducible. More precisely, it shows that if that assumption does not hold, the studied F-particle can be regarded as an F^0-particle, where F^0 is some vector space on which G operates irreducibly. It should be pointed out, however, that there is no way in general of defining F^0 *canonically*.

We have assumed that for any translation $\mathbf{a} \in \mathbf{E}_4$, there exists an element $g_{\mathbf{a}}$ of the structural group G which operates as this translation $\mathbf{a}$ on E_4 and operates identically on F. It will follow from the next theorem that this is always true in a very wide range of cases (an example of those is the case of the electron, where G is the proper spinor group).

Theorem. Let F be a finite-dimensional vector space over C and G be a topological group, operating irreducibly on F. Let G operate continuously on F, i.e., the map

$$(g, x) \in G \times F \to gx \in F$$

is continuous. Let G_0 be a subgroup of G having the following properties:

(1) G_0 is abelian and is an invariant subgroup of G.
(2) The only character on G_0, invariant through the connected component of the unity of G, is the character 1.

Then G_0 operates identically on F.

Remarks. Elements $\sigma \in G$ operate on G_0 by interior automorphisms $g_0 \in G_0 \to \sigma g_0 \sigma^{-1} \in G_0$. Also σ operates on the characters of G_0 by transfer of structure, i.e., if x is a character of G_0, then σx is a new character of G_0 defined by the equation

$$\sigma x(\sigma g_0 \sigma^{-1}) = x(g_0), \qquad g_0 \in G_0.$$

Let us show that for the electron, the hypotheses are satisfied. Here G is the proper spinor group, which is a two-order covering group of the proper inhomogeneous Lorentz group. The image of G into the affine group of E_4 is the proper inhomogeneous Lorentz group, but every point of the Lorentz group is the image of two different elements of G. G_0 is the connected component of the identity in the inverse image in G of the group of translations.

To prove condition (1) for the electron, first note that the group of translations is invariant and abelian in the proper inhomogeneous Lorentz group. The inverse image of an invariant subgroup is an invariant subgroup. Also, if a group is invariant, the connected component of the unity is invariant. Therefore G_0 is an invariant subgroup of G. On the other hand, G_0 is abelian because it is isomorphic with the group of translations.

To prove the second condition for the electron we shall first prove that the only character on the group of translations $\mathbf{E}_4$ invariant by the proper inhomogeneous Lorentz group is the character 1. The dual $\tilde{\mathbf{E}}'_4$ is canonically isomorphic to the space of the characters on $\mathbf{E}_4$. The proper inhomogeneous Lorentz group is connected, it acts on $\mathbf{E}_4$ by internal automorphisms simply as the proper homogeneous Lorentz group, and it acts on the characters as the proper homogeneous Lorentz group acts normally on the dual $\mathbf{E}'_4$. We only have to find the points of $\tilde{\mathbf{E}}'_4$ which are Lorentz invariant. But only the origin is invariant, and the origin defines the character 1. It follows that condition (2) is true for the electron because G_0 is isomorphic to the group of translations and the elements $\sigma \in G$ act on G_0 as the elements of the proper inhomogeneous Lorentz group act on $\mathbf{E}_4$.

Proof of the theorem. For any finite-dimensional complex representation of an abelian group there exists at least one common eigenvector; therefore there exists an eigenvector $\mathbf{f}$ for G_0:

$$g_0 \mathbf{f} = \chi(g_0)\mathbf{f} \text{ whatever be } g_0 \in G_0,$$

$\chi(g_0)$ being a complex function of $g_0 \in G_0$. Since the representation of G is continuous, the scalar $\chi(g_0)$ defines a continuous function of G_0, which is a character because

$$\chi(g_0 g_1) = \chi(g_0)\chi(g_1).$$

Thus for every eigenvector **f** we have an associated character χ. Several eigenvectors may correspond to the same character χ. Let $F_\chi \subset F$ be the maximal subspace formed of eigenvectors such that G_0 operates on F_χ as multiplication by the character χ. Let $\{F_i\}$ be the collection of those subspaces F_χ, when χ runs over the set of characters of G_0. The F_i's are independent. It is sufficient to show that for any minimal dependent system of vectors (which is finite because F is finite-dimensional), each of which corresponds to a character, the characters are identical. Let $e_1, e_2, \ldots, e_k$ be a minimal dependent system of vectors, each of which corresponds to a character χ_i. Any $l - 1$ of the e_i's is independent, and we have

$$\lambda_1 e_2 + \lambda_2 e_2 + \ldots + \lambda_k e_k = 0$$

for some set of non zero complex numbers $\{\lambda_i\}$. By the operation $g_0 \in G_0$, we have

$$\chi_1(g_0)\lambda_1 e_1 + \chi_2(g_0)\lambda_2 e_2 + \ldots + \chi_k(g_0)\lambda_k e_k = 0.$$

If $\chi_i(g_0)$ depends on i, the two equations can be combined to give a relation between a smaller number of vectors. But this is impossible. Therefore the characters $\chi_1, \chi_2, \ldots, \chi_k$ take the same value on g_0, and since this is true for every g_0, these characters are identical. The independence of the F_i's assures their number is finite.

Each $\sigma \in G$ operates on F and on G_0; since G_0 is an invariant subgroup of G, σ leaves invariant (as a set) the operation of G_0 on F. Hence σ must leave the collection $\{F_i\}$ unchanged (but the corresponding characters χ_i might be interchanged). If an element of G varies in the connected component of the unity it operates continuously on this finite set of χ_i; therefore it cannot interchange them and it leaves each χ_i invariant. But it was assumed that the only character on G_0 invariant through the connected component of the unity is 1. Therefore there exists only one maximal subspace F_1 such that G operates on F_1 by multiplication by a character χ_1, and this character χ_1 must be 1. Now G leaves F_1 invariant. Since the representation of G in F is irreducible, we must have $F_1 = F$. Therefore G_0 operates identically on F.

Determination of all Vector Particles

Let us summarize the problem. We have $V = E_4$, a finite-dimensional vector space F over C and a Lie group G satisfying the hypotheses:

(1) G preserves the Lebesgue measure on E_4, i.e., all $\sigma \in G$ have determinant ± 1 as operators on $\mathbf{E}_4$.
(2) For each translation on E_4 there exists at least one element of G that operates as this translation on E_4 and operates identically on F.

We are looking for Hilbert spaces $\mathscr{H} \subset \mathscr{D}'(E_4; F)$ with finer topology than the induced one, extremal, and G-invariant. As we have seen, it is equivalent to look for a measure $\boldsymbol{\mu}$ in $\tilde{\mathbf{E}}'_4$ with values in $F \otimes \overline{F}$, tempered, positive, G-invariant, and extremal. We have found that a necessary condition for $\boldsymbol{\mu}$ to be extremal is that the support of $\boldsymbol{\mu}$ must be in the closure of one G-orbit. For the meson we took $G =$ proper inhomogeneous Lorentz group and $F = C$. We found a scalar measure μ which was tempered, positive, G-invariant, and extremal.

Those measures which correspond to physically acceptable particles have as support one branch of an hyperboloid of two sheets (or one branch of the cone in the case of mass 0), and thus it depends on the parameters m_0, $\pm$. The vector measure $\boldsymbol{\mu}$ will also depend on the parameters m_0, $\pm$, and others in addition. We shall prove for the vector particles that if we remain on the hyperboloid of two sheets (i.e., $m_0 \neq 0$), the measure $\boldsymbol{\mu}$ must be of the form $\boldsymbol{\mu} = \mathbf{M}\mu$, where μ is the measure we found for the meson, and $\mathbf{M}$ is an infinitely differentiable function on the hyperboloid, having values in $F \otimes \overline{F}$, slowly increasing at infinity, G-invariant, and extremal.

Let W be a C^∞ manifold, G a Lie group that operates on W. We assume that $(g, x) \to gx$ is a C^∞ mapping $G \times W \to W$. We shall show that every G-orbit is a manifold in some local sense which we shall make more precise. Later we shall need the hypothesis that the G-orbits to be discussed are closed manifolds.

Consider the operation $s \to sa$, where $s \in G$, $a \in W$, and $sa \in W$. This operation is a C^∞ map, therefore it has a tangent map $\mathbf{X} \to \mathbf{Xa}$, where $\mathbf{X}$ is a tangent vector at the point s of the group and $\mathbf{Xa}$ is a tangent vector at the point sa of the manifold. In particular if $\mathbf{X}$ is an element of the Lie algebra $\mathfrak{G}$ (i.e., the tangent space to the group at the unit element), then $\mathbf{Xa}$ will be a tangent vector at the point a itself. Thus we see that every element of $\mathfrak{G}$ defines a vector field on W by this map (when a varies on W). We have thus a linear map from the Lie algebra $\mathfrak{G}$ into the vector space of all vector fields on W.

For a given $a \in W$, the range of the map $s \to sa$ is exactly the orbit of the point a.

Let $\mathfrak{M}, \mathfrak{M}'$ be two C^∞ manifolds, u a C^∞ mapping from $\mathfrak{M}$ into $\mathfrak{M}'$,

t_u the tangent map. For every point m of $\mathfrak{M}$, the restriction of t_u to the tangent space $T(m)$ to $\mathfrak{M}$ at m is a linear map from $T(m)$ into the tangent space $T'(m')$ to $\mathfrak{M}'$ at $m' = u(m)$. The dimension of the image of $T(m)$ under this linear map is called the *rank of the mapping u at the point m.*

We shall apply this definition to $\mathfrak{M} = G$, $\mathfrak{M}' = W$ and to the map $s \to sa$, when $a \in W$ is fixed.

Theorem. The map $s \to sa$ is of constant rank.

Proof. Any point $s_0 \in G$ defines a left translation of G, in particular an isomorphism of G onto itself. Hence the tangent map associated with such a translation is, at every point, a linear isomorphism (onto) of the corresponding tangent spaces. In particular, the map $\mathbf{Y} \to s_0\mathbf{Y}$ is a linear isomorphism of the Lie algebra $\mathfrak{G}$ onto the tangent space of G at s_0. Any $\mathbf{X}$ in the tangent space to G at s_0 may be written in the form $\mathbf{X} \to s_0\mathbf{Y}$, where $\mathbf{Y}$ is an element of the Lie algebra. Since G operates as a group on W, we have

$$(s_0 y)a = s_0(ya)$$

for s_0, $y \in G$. By differentiation, it follows that this associative law also holds for the tangent vectors:

$$(s_0\mathbf{Y})a = s_0(\mathbf{Y}a).$$

Thus we have

$$\mathbf{X}a = (s_0\mathbf{Y})a = s_0(\mathbf{Y}a).$$

The rank of the map $s \to sa$ at s_0 is the dimension of the vector subspace

$$A_{s_0} = \{\mathbf{X}a : \mathbf{X} \in \text{tangent space of } G \text{ at } s_0\}$$
$$= \{s_0(\mathbf{Y}a) : \mathbf{Y} \in \mathfrak{G}\}.$$

The dimension of

$$A_e = \{\mathbf{Y}a : \mathbf{Y} \in \mathfrak{G}\}$$

is the rank of the map $s \to sa$ at $s = e$, the unit element of G. The space A_{s_0} is the transform by t_{s_0} (tangent mapping associated to $s_0 : W \to W$) of the space A_e. Since s_0 is an automorphism of the manifold W, t_{s_0} defines linear isomorphisms (onto) of the tangent spaces. Therefore $\dim A_{s_0} = \dim A_e$.

The set G_a of elements $s \in G$ such that $sa = a$ ($a \in W$ fixed) is a subgroup of G, called *the stabilizer of a.*

For fixed $a \in W$, let p be the rank of the map $s \to sa$. The tangent map $\mathbf{X} \to \mathbf{X}a$, restricted to the tangent space to G at an arbitrary point s_0, is a linear map with rank p. Take $s_0 = e$, unit element of G. By the classical theorem of constant rank there exists a neighbourhood $\mathscr{E}$ of e in G such that the set $\{sa : s \in \mathscr{E}\}$ is exactly a C^∞-manifold of dimension p. Moreover $G_a \cap \mathscr{E}$ is also a manifold having a dimension which is the difference between the dimension of the group G and the dimension of the orbit. Therefore G_a is a true Lie subgroup of G. The tangent vector space to $G_a \cap \mathscr{E}$ at e, which is the Lie algebra $\mathfrak{G}_a$ of the stabilizer G_a, is exactly the set of all $\mathbf{X}$ having the image $\mathbf{X}a = 0$.

But it could happen that the orbit passes near the point a infinitely many times, each time closer to a, in such a way that the intersection of any sufficiently small neighbourhood of a, *in* W, with the orbit has always an infinite collection of sheets. In that case, the orbit cannot be submanifold of W.

In order to avoid such accidents, we make the hypothesis that the orbit is a *closed submanifold* of W. Then the dimension of the manifold is the constant p, because it is a countable union of pieces of manifold of dimension p. When G = Lorentz group or spinor group and $W = E_4'$, the orbits we shall consider will be the pure hyperboloids, the light cone itself being excluded by our hypotheses that the orbits must be closed.

Now we shall use the theorem of implicit functions to obtain more results. Consider a transversal manifold Σ through the unit $e \in G$, i.e., a C^∞-manifold having a tangent space supplementary to the tangent space of the stabilizer G_a. (Two vector subspaces are supplementary if any vector of the space is in one and only one way the sum of two vectors belonging respectively to those subspaces.) This transversal manifold Σ implies the choice of a subspace in $\mathfrak{G}$, supplementary

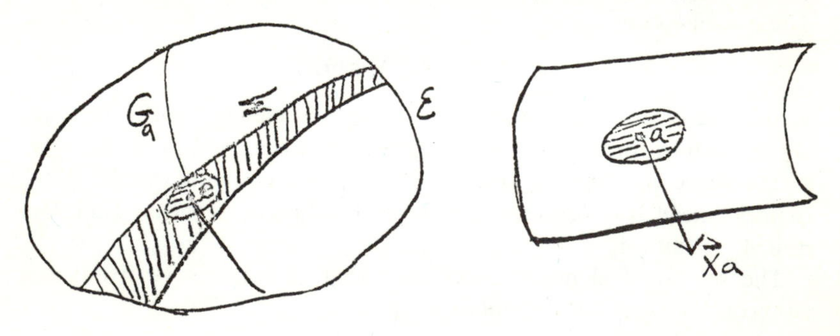

to $\mathfrak{G}_a$. We may restrict our considerations to Σ because $a \in W$ is invariant under the operation of $s \in G_a$. We may obtain all of the intersection of the orbit with a neighbourhood of a by considering only the operation of Σ on a.

Let Ω be the orbit of a and consider the equation

$$y = sx \text{ for } x, y \in \Omega, s \in \Sigma.$$

When we regard y as a function of s, we see that the Jacobian matrix of y with respect to s (in admissible local coordinates for s near e and for x, y near a in Ω) is nothing else but the matrix (in the basis defined by those local coordinates) of the tangent mapping. For $s = e$, $x = a$, this tangent map is an isomorphism of the tangent space to Σ at e onto the tangent space to Ω at a. Therefore the Jacobian cannot be zero. Because of this fact, we may apply the following result.

Implicit Function Theorem.

There exists a neighbourhood X' of a in Ω and a neighbourhood Σ' of e in Σ such that for all $x \in X'$, $y \in X'$, the equation $y = sx$ has one and only one solution $s \in \Sigma'$. This solution may be written $s = S(x, y)$, and S is a C^∞-function $X' \times X' \to \Sigma'$. (Furthermore, because of the continuity of the operation $(s, x) \to sx$, there exists neighbourhoods X'' of a and Σ'' of s with $\overline{X''} \subset X'$ such that for $x \in X''$, $s \in \Sigma''$, we have $sx \in X'$. We may go one step further and find neighbourhoods X_0 of a and Σ_0 of e with $X_0 \subset X''$ such that for $x \in X_0$, $s \in \Sigma_0$ we have $s^{-1} \in X''$.)

Theorem. Let $\mathbf{T}$ be a distribution on the manifold Ω with values in $F \otimes \overline{F}$. If $\mathbf{T}$ is G-invariant, then $\mathbf{T}$ is an infinitely differentiable function.

Remarks. In this statement, $F \otimes \overline{F}$ can be replaced by any finite-dimensional vector space.

We recall what one means by saying that $\mathbf{T}$ is a C^∞ function on Ω. Take any open subset U of Ω on which there are local coordinates. By definition, those coordinates define a C^∞-homeomorphism of U onto an open subset U' of R^p (p = dimension of Ω). Hence we may carry the restriction of $\mathbf{T}$ to U into a distribution $\mathbf{T}'$ on U' (still valued in $F \otimes \overline{F}$). Now this distribution $\mathbf{T}'$ on U' is a C^∞ function on U', that is to say, if we choose a Lebesgue measure dx in R^p there is a C^∞-function $\mathbf{f}(x)$ on U' valued in $F \otimes \overline{F}$ such that

$$\langle \mathbf{T}', \varphi \rangle = \int \mathbf{f}(x)\varphi(x)\, dx,$$

for any $\varphi \in \mathscr{D}(R^p)$ with support in U'. This property does not depend on the choice of U or on the local coordinates on U, nor on the Lebesgue measure dx.

Proof of the theorem. Let Φ be the representation of G in $F \otimes \overline{F}$. From the invariance of $\mathbf{T}$, we have

$$\langle \mathbf{T}, \varphi \rangle = \Phi(s) \langle \mathbf{T}, \varphi(sx) \rangle, \; \varphi \in \mathscr{D}(\Omega).$$

We have to prove that for every $a \in \Omega$, $\mathbf{T}$ is infinitely differentiable in some neighbourhood of a. It is sufficient that this neighbourhood be any X_0 chosen as in the Implicit Function Theorem. Let $\varphi \in \mathscr{D}(X_0)$, $\alpha \in \mathscr{D}(\Sigma_0)$. Let us set $ds = ds_1 ds_2 \ldots ds_p$, where the s_i are canonical coordinates on the local map Σ_0. If $\int \alpha(s)\, ds = 1$, $\langle T, \varphi \rangle = \int \alpha(s)\, ds\, \Phi(s) \times \langle T, \varphi(sx) \rangle$. From the definition of tensor products of distributions,

$$\begin{aligned} \langle T, \varphi \rangle &= \langle \alpha(s)\Phi(s) \otimes T_x, \varphi(sx) \rangle \\ &= \langle T_x, \textstyle\int \alpha(s)\Phi(s)\, \varphi(sx)\, ds \rangle. \end{aligned}$$

To compute the integral we need to consider only those values of x where $\varphi(sx)$ is different from zero, i.e., the set

$$\{x : sx \in \text{support of } \varphi, s \in \text{support of } \alpha\}.$$

But the support of φ is contained in X_0 and the support of α is contained in Σ_0. By setting $sx = y$ we find that it is sufficient to consider x in the set

$$\{x = s^{-1}y : y \in X_0, s \in \Sigma_0\}.$$

But by the Implicit Function Theorem this set is contained in X''; thus it is sufficient to consider $x \in X''$: for $x \notin X''$, the integrand is zero. Let us use the equation $s = S(x, y)$ to make a change of variables from the old variable s to a new variable y (with x fixed). We must check that the domains are well defined for the old and the new variables and that the change of variables defines an isomorphism between these two domains with a Jacobian different from zero. The map $y \to s = S(x, y)$ for x fixed in X'' is a map from X' into Σ'. Since this map is inversible by $y = sx$, this map carries X' into an open subset of Σ'. Thus this map defines a C^∞-homeomorphism of X' into an open subset of Σ'. Now we must show that the image of X' covers the whole domain of integration with respect to s, i.e., we must show that the support of $\alpha(s)$ is contained in the image of X'. If s belongs to the support of $\alpha(s)$, then $s \in \Sigma_0$. But $x \in X''$, therefore $sx \in X'$, and it follows that the image of X' under the map $s = S(x, y)$ for x fixed in X'' covers Σ_0, and therefore it also covers the support of $\alpha(s)$. The change of variables gives:

$$\begin{aligned}\langle T, \varphi\rangle &= \langle T_x, \int \alpha(S(x, y))\, \Phi(S(x, y))\, \varphi(y) |\det (\partial S/\partial y)(x, y)|\, dy\rangle \\ &= \int \varphi(y)\, dy\, \langle T_x, \alpha(S(x, y))\, \Phi(S(x, y)) |\det (\partial S/\partial y)(x, y)|\rangle.\end{aligned}$$

Now

$$\alpha(S(x, y))\, \Phi(S(x, y)) |\det (\partial S/\partial y)(x, y)|$$

is an infinitely differentiable function of y in X'. It follows that

$$\tau(y) = \langle T_x, \alpha(S(x, y))\, \Phi(S(x, y)) |\det (\partial S/\partial y)(x, y)|\rangle$$

is also an infinitely differentiable function of y. We now have

$$\langle T, \varphi\rangle = \int \varphi(y)\, \tau(y)\, dy = \langle \tau, \varphi\rangle$$

therefore $T = \tau$, an infinitely differentiable function of y.

Let us apply this theorem to our problem. We have a G-orbit in Ω, which is an hyperboloid, and we have a measure $\boldsymbol{\mu}$ with support Ω, which defines a distribution on Ω. By the theorem, we know that $\boldsymbol{\mu}$ is an infinitely differentiable function on Ω. The same result for the scalar measure μ is true, i.e., it is an infinitely differentiable function. By the invariance of μ, it follows that μ is never zero, otherwise it would be zero everywhere. We may take the quotient $\boldsymbol{\mu}/\mu = \mathbf{M}$ and $\mathbf{M}$ is an infinitely differentiable function on Ω. Therefore we have the result that $\mathbf{M}$ may be written in the form $\boldsymbol{\mu} = \mathbf{M}\mu$, where $\mathbf{M}$ is an infinitely differentiable function. Moreover, $\mathbf{M}$ is invariant by G because if $\sigma \in G$ acts on it, $\mathbf{M}\mu$ is transformed into another distribution which defines the same measure; therefore $\mathbf{M}$ must be replaced by another function which is μ-almost everywhere equal to $\mathbf{M}$, but since $\mathbf{M}$ is continuous, $\mathbf{M}$ must be equal to the transformed function. If $\boldsymbol{\mu}$ is positive, $\mathbf{M}$ is positive because it is μ-almost everywhere positive, and since $\mathbf{M}$ is continuous, it is everywhere positive.

The problem has been reduced to the search for an infinitely differentiable function $\mathbf{M}$ on the hyperboloid Ω, having values in $F \otimes \overline{F}$, G-invariant, positive, extremal, and slowly increasing at infinity. Take a point $a \in \Omega$. We know that $\mathbf{M}(a)$ is a positive G_a-invariant element of $F \otimes \overline{F}$. Conversely, if we let $M(a)$ be any positive G_a-invariant element of $F \otimes \overline{F}$ then there corresponds to it in a unique way an infinitely differentiable function $\mathbf{M}$ on Ω, having values in $F \otimes \bar{F}$, G-invariant, positive, extremal, and slowly increasing. It is given by

$$\mathbf{M}(p) = \sigma M(a)$$

for $p = \sigma a$; $\mathbf{M}(p)$ is independent of σ because if another element σ' carries a into p, it has the form $\sigma' = \sigma\gamma$, where $\gamma \in G_a$, and it follows that

$$\sigma' M(a) = \sigma\gamma M(a) = \sigma M(a)$$

since $M(a)$ is invariant by G_a. We must prove that $M(p)$ is infinitely differentiable. Consider a neighbourhood N_a of a in Ω sufficiently small to apply the Implicit Function Theorem. Then we may take $\sigma = S(a, p)$, where S is determined by the Implicit Function Theorem. Thus

$$\mathbf{M}(p) = S(a, p) M(a)$$

and since $S(a, p)$ is a C^∞-function, we see that $M(p)$ is a C^∞-function in the neighbourhood N_a of a. Now if we take any point b, there is a σ_0 such that $b = \sigma_0 a$. The set $N_b = \{\sigma_0 p : p \in N_a\}$ is a neighbourhood of b. Since any $p \in N_a$ can be written $p = \sigma a$ with $\sigma = S(a, p)$, it follows that any $q \in N_b$ can be written $q = \sigma_0 S(a, p) a = \sigma_0 S(a, \sigma_0^{-1} q) a$, and then for $q \in N_b$ we have

$$\mathbf{M}(q) = \sigma_0 S(a, \sigma_0^{-1} q) \,.\, M(a).$$

Thus $\mathbf{M}(q)$ is C^∞ for $q \in N_b$. It follows then that $\mathbf{M}$ is infinitely differentiable on all of Ω.

We have yet to prove that M defined in this way is slowly increasing at infinity. Assume that G is exactly the proper homogeneous Lorentz group. Consider a coordinate system in $\tilde{\mathbf{E}}_4$ in which a has the form $a = (0, 0, 0, a_0)$. In the equation

$$\mathbf{M}(p) = \sigma \,.\, M(a),\ p = \sigma a,$$

the operation of σ on $M(a)$ is precisely given by $\Phi(\sigma)$. Thus for clarity it is better to write

$$\mathbf{M}(p) = \Phi(\sigma) \,.\, M(a),\ p = \sigma a.$$

Take any norm in $F \otimes \bar{F}$. Then

$$\|\mathbf{M}(p)\| \leqslant \|\Phi(\sigma)\|\ \|M(a)\|,\ p = \sigma a.$$

Now it is sufficient to show that $\|\Phi(\sigma)\|$ is slowly increasing with respect to p where $p = \sigma a$. Assume that p has the form

$$p = (p_1, 0, 0, p_0)$$

and consider the Lorentz transformations (θ), depending on the parameter θ, in the plane P of the first space axis and the time axis:

$$(\theta)\colon \begin{cases} p_1 = q_1 \cosh \theta + q_0 \sinh \theta \\ p_0 = q_1 \sinh \theta + q_0 \cosh \theta \end{cases}$$

Since we start from the point $(0, 0, 0, a_0)$, we have

$$p_1 = a_0 \sinh \theta$$

$$p_0 = a_0 \cosh \theta.$$

Therefore

$$p_1 \sim \tfrac{1}{2} a_0 e^{\theta}$$

$$p_0 \sim \tfrac{1}{2} a_0 e^{\theta},$$

and thus for some constant C_1, we have

$$|\theta| \leqslant \log(C_1 \|p\|.)$$

Now let p be an arbitrary point on the hyperboloid. We may pass from a to p by two successive Lorentz transformations. Let τ be a space rotation φ (which is an element of the stabilizer G_a of a), which puts p in the plane P; let us set $p' = \tau p$. Let (θ) be a transformation in the plane P as described, sending a into p'. We have: $\sigma = \tau^{-1} . (\theta)$ and

$$\|\Phi(\sigma)\| \leqslant \|\Phi(\tau^{-1})\| \, \|\Phi((\theta))\|.$$

But the space rotations form a compact group of linear operators in $F \otimes \overline{F}$. Therefore all their norms are bounded by some fixed constant. Then for some constant C_2,

$$\|\Phi(\sigma)\| \leqslant C_2 \|\Phi((\theta))\|.$$

By a general law about the representation of any one parameter group in a normed space which states that the norm is bounded by an exponential of the parameter, we obtain the result

$$\|\Phi((\theta))\| \leqslant \exp[k|\theta|]$$

for some number k. Since $|\theta| \leqslant \log C_1 \|p'\|$, we may write

$$\| \Phi((\theta)) \| \leqslant C_3 \exp [k \log (C_1 \| p' \|)] = C_4 \| p' \|^k.$$

But again, since the space rotations form a compact group, we may estimate $\|p'\| \leqslant C_5 \|p\|$, and finally for some constant C_6, we have

$$\|\Phi((\theta))\| \leqslant C_6 \|p\|^k;$$

hence $\|\Phi(\sigma)\|$ is tempered and $\mathbf{M}$ is slowly increasing at infinity.

If G is any group such that its image in the linear group of $\tilde{\mathbf{E}}'_4$ is the proper homogeneous Lorentz group, a one-parameter group (θ) in the image may be raised into a one-parameter group in G. Simply raise the infinitesimal generator and it will generate a one-parameter group in G which has (θ) as its projection. Thus the proof of the slowly increasing property of $\mathbf{M}$ is valid for such a group G.

Now we have shown that all the G-invariant Hilbert spaces $\mathscr{H} \subset \mathscr{D}'(E_4; F)$ (continuous injection), relative to the orbit Ω (i.e., with parameters $m_0 > 0$ and $\pm$), are in one-to-one correspondence with the elements $M(a) \in F \otimes \overline{F}$ for fixed $a \in \Omega$, positive, and G_a-invariant. In addition $\mathscr{H}$ is extremal if and only if the corresponding $M(a)$ is extremal.

We may now apply the theory of kernels. An element of $F \otimes \overline{F}$, positive, and G_a-invariant is a positive G_a-invariant anti-kernel relative to F, i.e., a positive anti-linear map $F' \to F$ which is G_a-invariant. Finding this anti-kernel is equivalent to finding a subspace $F_a \subset F$, equipped with an hermitian structure and G_a-invariant. This subspace is extremal if and only if the corresponding element of $F \otimes \overline{F}$ is extremal.

The methods for finding all the G_a-invariant and irreducible subspaces F_a of F are known. Assume that we have found such an F_a, we must put a G_a-invariant hermitian structure on F_a if possible. Let Γ_a be the image of G_a in the linear group of F_a. Note that $\Gamma_a = G_a/G_1$, where G_1 is the subgroup of G_a which operates identically on F_a. The problem is then to put a Γ_a-invariant hermitian structure on F_a. The necessary and sufficient condition for its existence is that Γ_a be relatively compact. To prove this, first suppose that Γ_a leaves a hermitian form over F_a invariant. Then Γ_a must be contained in the unitary group of this hermitian structure. Since the unitary group is compact, Γ_a is relatively compact. Conversely, if Γ_a is relatively compact we may choose any hermitian form on F_a, and its average with respect to the Haar measure of $\overline{\Gamma}_a$ is a Γ_a-invariant hermitian form. This Γ_a-invariant hermitian form is unique to a constant factor.

All the possible solutions for a given orbit Ω have now been determined. Let us summarize the procedure. We take in F all the possible subspaces F_a which are G_a-invariant, and G_a-irreducible. For each such F_a, if the image Γ_a of G_a in the linear group of F_a is relatively compact, then there exists one and only one hermitian form defined to a constant factor which is G_a-invariant, and it gives a vector particle. If the image Γ_a is not relatively compact, then we have found no particle.

As an example let G = proper inhomogeneous Lorentz group, $V = E_4$, F = any finite dimensional vector space. Let Ω be any sheet of a hyperboloid of two sheets (but not the cone) in $\mathbf{E}'_4$. Choose a point $a \in \Omega$ and choose a coordinate system such that the point a has coordinates $(0, 0, 0, a_0)$. The stabilizer G_a is simply the orthogonal group in the three-dimensional space-like plane defined by the three space axes of this chosen coordinate system. We know that F is completely reducible. To each irreducible subspace of F will correspond an F-particle because the orthogonal group is compact. If, in these irreducible subspaces, the representations of G_a are not equivalent, this decomposition is unique. But there might be several equivalent representations, and in this case we have an infinity of possible irreducible subspaces. Every one gives an F-particle. We are not concerned here, however, with the problem of determining whether two reducible equivalent representations give the same particles or not; we are only looking for F-particles for one given representation. Our problem of determining all F-particles is now solved for G = proper inhomogenous Lorentz group, and it is easily seen that the problem for G = proper spinor group is solved in a similar way.

Now if we consider a hyperboloid of one sheet the problem is quite different. Given the point a on this hyperboloid, choose a coordinate system such that the coordinates of a are $(a_1, 0, 0, 0)$. The stabilizer G_a is the Lorentz group in the last two space coordinates and the time coordinate. We have to see if F has any irreducible components F_a on which G_a acts through a compact factor (that is, the image of G_a in the linear group of F_a is a compact group). It may happen, for instance, that G_a acts on F_a as the identity. This is what happens in the case of the meson, where $F = C$ and the whole group G acts on C identically (and, of course, G_a also acts identically on C). It turns out that the only F-particles we obtain are the ones where G_a acts on F_a as the identity because the only compact factor group of the Lorentz group is the identity.

The subspace $\mathbf{F}_a$ *when* $m = 0$. For $m = 0$ the hyperboloid is replaced by the light cone. For any point a on the light cone (except the origin), the stabilizer G_a is the Lorentz group of the tangent plane, which is isomorphic to the group of rotations and translations in a two-dimensional Euclidean plane. It has as a compact factor the orthogonal group in the two-dimensional plane. Thus, for some representations we shall have an $M(a)$ and for others we shall not. In the case of photons it is not the case. It is known in physics that photons are not represented by Hilbert spaces of functions but by non-separate pre-Hilbert complete spaces. In such a space, a motion is a class of ψ functions equivalent with respect to gauge invariance.

Complete Description of $\mathscr{F}\mathscr{H}$ for Vector Particles

First we shall describe a functional space and prove later that it is $\mathscr{F}\mathscr{H}$. Let a be a given point on one sheet Ω of a hyperboloid with two sheets. We have the scalar measure μ with support Ω, and the function $M(p)$ defined on Ω. For any given point $p \in \Omega$, $M(p)$ is a positive element of $F \otimes \overline{F}$, that is, it is a positive anti-linear map (positive anti-kernel):

$$F' \to F_p \subset F$$

where the subspace F_p has a G_p-invariant hermitian structure, is G_p-irreducible, and corresponds to this anti-kernel $M(p)$ in the Theorem of Anti-kernels. Since $M(p) = \sigma M(a)$ for any σ satisfying $p = \sigma a$, then $F_p = \sigma F_a$ for any such σ, where σF_a means the action of σ on F_a in the representation of G into F. The quadratic form of F_p is the transform by σ of the quadratic form on F_a. Thus for each $p \in \Omega$, we have a subspace $F_p \subset F$ which varies with p and has a positive definite hermitian form which also varies with p.

Definition. $L^2(\Omega, \mu, F_p, F)$ is the space of classes of functions f (μ-almost everywhere equal):
$\Omega \to F$ with the following properties:

(1) for each $p \in \Omega, f(p) \in F_p$;
(2) as function valued in F, f is μ-measurable;
(3) $\|f(p)\|_p$ is square μ-integrable.

We denote by $(,)_p$ the hermitian form in F_p and

$$\|\mathbf{f}\|_p = (\mathbf{f}|\mathbf{f})_p^{\frac{1}{2}}, \mathbf{f} \in F_p.$$

We provide $L^2(\Omega, \mu, F_p, F)$ with the inner product

$$\int_\Omega (f(p)/g(p))_p \, d\mu(p), f, g \in L^2(\Omega, \mu, F_p, F).$$

The associated norm is

$$\|f\|_{L^2} = (\int_\Omega \|f(p)\|_p^2 \, d\mu(p))^{\frac{1}{2}}$$

Lemma. If f is a function $\Omega \to F$ satisfying the conditions (1) and (2) of the above definition, the non-negative function $\|f(p)\|_p$ is μ-measurable.

Let $a \in \Omega$ be fixed. There is a neighbourhood Ω_a of a in Ω such that, for $p \in \Omega_a$, we may consider the function $S(a, p)$ of p, valued in G, as defined by the Implicit Function Theorem; $S(a, p)$ is a C^∞ function $\Omega_a \to G$.

We have:

$$\|f(p)\|_p = \|S^{-1}(a, p)f(p)\|_a, p \in \Omega_a.$$

If $f(p)$ is μ-measurable, as function $\Omega_a \to F$, the same is true of $S^{-1}(a, p)$ $f(p)$, as function $\Omega_a \to F$ and also as function $\Omega_a \to F_a$. Whence the lemma: for the norm of measurable vector function (with values in a finite-dimensional space) is measurable; and because measurability is a local property.

Theorem. *$L^2(\Omega, \mu, F_p, F)$ is an Hilbert space.* Let $\{f_k\}$ be a Cauchy sequence of elements of $L^2(\Omega, \mu, F_p, F)$. We can find a subsequence $\{f_{k_\alpha}\}$ ($\alpha = 1, 2, \ldots$) such that

$$\|f_{k_{\alpha+1}} - f_{k_\alpha}\|_{L^2} \leqslant 2^{-\alpha}$$

Let us set $g_\alpha = f_{k_{\alpha+1}} - f_{k_\alpha}$. If we prove that the series $\sum_{\alpha=1}^{\infty} g_\alpha$ converges in $L^2(\Omega, \mu, F_p, F)$ to an element g of that space, then $f = g + f_{k_1}$ will be the limit of the *sequence* $\{f_k\}$.

The series

$$\sum_{\alpha=1}^{\infty}\left(\int_\Omega \|g_\alpha(p)\|_p^2 \, d\mu(p)\right)^{\frac{1}{2}}$$

converges. From a property of scalar functions, we conclude that there is a set $N \subset \Omega$ of measure zero such that the series

$$\sum_{\alpha=1}^{\infty} \|g_\alpha(p)\|_p$$

converges if $p \notin N$. But that implies that, for $p \notin N$, the series $\sum_{\alpha=1}^{\infty} g_\alpha(p)$ converges in F_p; let $g(p)$ be its sum. For $p \in N$, let us set $g(p) = 0$. The series $\sum_{\alpha=1}^{\infty} g_\alpha$ converges μ-almost everywhere to the class of g (which we still denote by g). Since the g_α are measurable "functions" $\Omega \to F$, the same is true of g, according to Egoroff's theorem. According to the lemma, $\|g(p)\|_p$ is then measurable; and by applying a classical result for functions $\geqslant 0$

$$\Big(\int_\Omega \|g(p)\|_p^2 \, d\mu(p)\Big)^{\frac{1}{2}} \leqslant \sum_{\alpha=1}^{\infty} \Big(\int_\Omega \|g_\alpha(p)\|_p^2 \, d\mu(p)\Big)^{\frac{1}{2}}.$$

Hence g satisfies conditions (1), (2), (3) of the Definition. Q.E.D.

Definition. $\Lambda^2(\Omega, \mu, F_p, F)$ is the subspace of $\mathscr{D}'(E'_4) \otimes F$ of all measures $f\mu$ on $\mathbf{E}'_4$, where $f \in L^2(\Omega, \mu, F_p, F)$. $\Lambda^2(\Omega, \mu, F_p, F)$ is provided with the norm

$$\|f\mu\|_{\Lambda^2} = \|f\|_{L^2}.$$

Let $\mathscr{D}(\Omega, F_p, F)$ be the space of functions $\varphi: \Omega \to F$ with the following properties:

(1) for every $p \in \Omega$, $\varphi(p) \in F_p$;
(2) as function $\Omega \to F$, φ is a C^∞ function with compact support.

Theorem. $\mathscr{D}(\Omega, F_p, F)$ is a dense subspace of $L^2(\Omega, \mu, F_p, F)$. Likewise, $\{\varphi\mu;\ \varphi \in \mathscr{D}(\Omega, F_p, F)\}$ is a dense subspace of $\Lambda^2(\Omega, \mu, F_p, F)$.

Proof. Any function $f \in L^2 = L^2(\Omega, \mu, F_p, F)$ is the limit of truncated functions; that is, of functions equal to f on some compact subset of Ω and to zero elsewhere (one takes a sequence of such compact subsets increasing and converging to Ω). Therefore it is sufficient to prove that any $f \in L^2$ with compact support is the limit of functions in $\mathscr{D}(\Omega, \mu, F_p, F)$. In fact, it is sufficient just to prove that for any $a \in \Omega$ there is a neighbourhood of a such that any function having a compact support contained in this neighbourhood is the limit of functions in $\mathscr{D}(\Omega, \mu, F_p, F)$. If this is proved, then any compact support can be covered by a finite number of these neighbourhoods, and by a partition of unity it will follow immediately that any function $\in L^2$ having compact support will be the limit of functions of $\mathscr{D}(\Omega, \mu, F_p, F)$. Now, given $a \in \Omega$, we can choose an open neighbourhood Ω_a according to the Implicit Function Theorem, in such a way that for any function f on Ω_a, we may consider the function $g(p) = S^{-1}(a, p)f(p)$. Thus, in Ω_a, we are led back to a fixed Hilbert space F_a with the fixed norm $\|\cdot\|_a$. But it is known that the functions $\psi \in \mathscr{D}(\Omega_a, F_a)$ are dense in the Hilbert space $L^2(\Omega_a, \mu, F_a)$.

Therefore, the functions $p \to \varphi(p) = S(a, p)\ \psi(p)$ are dense in $L^2(\Omega_a, \mu, F_p, F)$. Moreover, φ is C^∞ because ψ and S are C^∞ and has a compact support since this is true of ψ. Q.E.D.

The space $\mathscr{H}$ is the completion of the space of all elements of the form $H * \overline{\varphi}$, $\varphi \in \mathscr{D}(E_4) \otimes F'$, for the norm

$$\|H * \overline{\varphi}\|_{\mathscr{H}} = \langle H * \overline{\varphi}, \varphi\rangle^{\frac{1}{2}}$$

We may also take $\varphi \in \mathscr{S}(E_4) \otimes F'$. The Fourier transform is

$$\mathscr{F}(H * \overline{\varphi}) = M\mu\tilde{\overline{\psi}}$$

where $\psi = \mathscr{F}\varphi$ and $\mathscr{F}H = M\mu$. Since $\psi = \overline{\mathscr{F}}\varphi$, it follows from the definition of the Fourier transform of distributions that

$$\begin{aligned}\|H * \overline{\varphi}\|^2_{\mathscr{H}} &= \langle H * \overline{\varphi}, \mathscr{F}\overline{\mathscr{F}}\varphi\rangle = \langle \mathscr{F}(H * \overline{\varphi}), \overline{\mathscr{F}}\varphi\rangle \\ &= \langle M\mu\tilde{\overline{\psi}}, \check{\psi}\rangle = \int_\Omega M\tilde{\overline{\psi}}\check{\psi}\,d\mu.\end{aligned}$$

But $M(p)$ defines an anti-linear map

$$\mathfrak{M}(p): F' \to F_p$$

given by the equation $\mathfrak{M}(p)\overline{g}, = M(p)\overline{\overline{g}}$, thus

$$M\mu\tilde{\overline{\psi}} = (\mathfrak{M}(p) \cdot \check{\psi}(p))\ \mu.$$

Let us set $f(p) = \mathfrak{M}(p) \cdot \check{\psi}(p)$ for $p \in \Omega$. We see that for each point $p \in \Omega, f(p) \in F_p$. If instead of an arbitrary $\varphi \in \mathscr{S}$, we take a $\varphi \in \mathscr{S}(E_4) \otimes F'$ such that its Fourier transform $\psi \in \mathscr{D}(E'_4) \otimes F'$, then we shall obtain an $f\mu$ where f is an infinitely differentiable function $\Omega \to F$

with compact support, taking its values at p in F_p. Thus $f\mu \in \Lambda^2(\Omega, \mu, F_p, F)$ and the norm of $f\mu$ in $\mathscr{F}\mathscr{H}$ is defined to be

$$\|f\mu\|_{\mathscr{F}\mathscr{H}} = \|H * \overline{\varphi}\|_{\mathscr{H}},$$

therefore:

$$\begin{aligned}\|f\mu\|^2_{\mathscr{F}\mathscr{H}} &= \int_\Omega M(p) \cdot \tilde{\psi}(p)\, \check{\psi}(p)\, d\mu(p) \\ &= \int_\Omega (\mathfrak{M}(p) \cdot \tilde{\psi}(p)) \cdot \check{\psi}(p)\, d\mu(p).\end{aligned}$$

But this last expression is of the form $\langle e', Le'\rangle$, and from the fact that $\langle e', Le'\rangle = \|Le'\|^2_{\mathscr{H}_1}$ where $\mathscr{H}_1$ is the Hilbert space corresponding to the anti-kernel L, it follows that

$$\int (\mathfrak{M}(p) \cdot \check{\psi}(p))\, \check{\psi}(p)\, d\mu(p)$$

is equal to the norm of f in

$$L^2(\Omega, \mu, F_p, F).$$

Hence:

$$\|f\mu\|^2_{\mathscr{F}\mathscr{H}} = \int \|f(p)\|^2_p\, d\mu(p),$$

and we have proved the

Theorem. If $f(p) = \mathfrak{M}(p) \cdot \check{\psi}(p)$ for $\psi \in \mathscr{D}(E'_4) \otimes F'$, then $f(p) \in \mathscr{D}(\Omega, F_p, F)$ and $\|f\mu\|_{\mathscr{F}\mathscr{H}} = \|f\mu\|_{\Lambda^2}$.

The space $\{f(p)\mu : f(p) = \mathfrak{M}(p) \cdot \check{\psi}(p), \psi \in \mathscr{D}(E'_4) \otimes F'\}$

with the Λ^2 norm is a dense subspace of $\mathscr{F}\mathscr{H}$. Now if we can show that this subspace is equal to $\{f(p)\mu\colon f(p)\mu \in \Lambda^2, f \in \mathscr{D}(\Omega, F_p, F)\}$, which is dense in $\Lambda^2 = \Lambda^2(\Omega, \mu, F_p, F)$, then it will follow that $\mathscr{F}\mathscr{H} = \Lambda^2$ because $\mathscr{F}\mathscr{H}$ will be the concrete completion (that is, the completion embedded in $\mathscr{D}'$) of a dense subspace of Λ^2. In order to show that these two sets are equal, we shall prove the following result:

Lemma. Whatever be $f \in \mathscr{D}(\Omega, F_p, F)$, there is $g \in \mathscr{D}(\Omega) \otimes F'$ such that $f(p) = \mathfrak{M}(p)\, g(p)$ for every $p \in \Omega$.

Proof of the theorem. Let a be any point of Ω and pick a basis of F', $\{e_1, \ldots, e_k, g_1, \ldots, g_l\}$, such that $\{\mathfrak{M}(a)e_1, \ldots, \mathfrak{M}(a)e_k\}$ is a basis of F_a. Then $\{\mathfrak{M}(p)e_1, \ldots, \mathfrak{M}(p)e_k\}$ is a basis of F_p for every $p \in \Omega$. We may write

$$f(p) = \sum_{i=1}^k f^i(p)\, \mathfrak{M}(p) e_i.$$

The components $f^i(p)$ of $f(p)$ are C^∞-functions on Ω with compact support. If we set

$$g(p) = \sum_{i=1}^{k} \overline{f^i(p)} e_i, \; p \in \Omega,$$

the conditions of the lemma are satisfied.

Now, if g is an infinitely differentiable function on Ω with compact support (valued in F'), it is easy to see that there is a continuation $\psi \in \mathscr{D}(E_4') \otimes F'$ such that $\psi(p) = g(p)$ for every $p \in \Omega$. This fact and the lemma prove that, whatever be $f \in \mathscr{D}(\Omega, F_p, F)$, there is $\psi \in \mathscr{D}(E_4') \otimes F'$, such that $f(p) = \mathfrak{M}(p)\check{\psi}(p)$ for every $p \in \Omega$. That proves:

Theorem. $\mathscr{F}\mathscr{H}$ is identical (as an Hilbert space) with $\Lambda^2(\Omega, \mu, F_p, F)$.

The Electron

We shall give a short description of the case of the electron. Here $F = C^2$ and G equals the covering group of the inhomogeneous proper Lorentz group. It can be shown that there is a C^∞ isomorphism of the two-order covering group of the homogeneous proper Lorentz group (that is, the proper spinor group) onto the unimodular group in C^2. The two elements of the spinor group which have the same projection in the Lorentz group correspond to two transformations in C^2 which differ only by the sign. The image of G into the affine group of E_4 is the inhomogeneous proper Lorentz group. The inverse image of the subgroup of the translations has two connected components. The connected component of unity operates identically on F and the other operates as -1 on F. Thus G does not operate faithfully on E_4 because two elements which have the same projection in the proper inhomogeneous Lorentz group give the same operator in E_4, and G does not operate faithfully on F because the translations operate identically. However, G operates faithfully on the product $E_4 \times F$.

Take a point a on the hyperboloid in $\mathbf{E}_4'$ corresponding to the mass m and consider the Lorentz reference frame in which the space-like components of a are zero. The stabilizer is then the orthogonal group in the three-dimensional subspace determined by the space-coordinate axes: G_a is the covering group of this orthogonal group; G_a is compact and F is G_a-irreducible. On F there is one and only one hermitian form (to a constant factor) which is G_a-invariant. Thus we may build the F_p as before and we will have the two-component function ψ_j

whose Fourier transforms are $\mathscr{F}\psi_j = f_j\mu, j = 1, 2$; we recall that $\psi = (\psi_1, \psi_2)$ and $\mathbf{f} = (f_1, f_2)$.

Then it can be proved that there exists one and only one (up to a constant factor) homogeneous first-order G-invariant linear differential operator $-\overline{D}$ with constant coefficients belonging to $\mathscr{L}(F, \overline{F})$. In our reference frame, we may write

$$-\overline{D} = \sum_{\mu} \sigma_\mu \partial_\mu$$

Similarly, there is one and only one (up to a constant factor) homogeneous first-order G-invariant linear differential operator $\overline{D}$ with constant coefficients in $\mathscr{L}(\overline{F}; F)$. We may consider the products $D\overline{D}$ and $\overline{D}D$. We shall normalize them in order to have:

$$\overline{D}D = \overline{J}D\,\overline{D}J = \square = m^2$$

on the wave functions of F-particles, where J is the canonical anti-isomorphism $F \to \overline{F}$ and $\overline{J}\,\overline{F} \to F$ its inverse.

If ψ is an F-particle wave function, $D\psi$ is an $\overline{F}$-particle (that is, an anti-particle) wave function. Similarly, if ψ is an $\overline{F}$-particle, $\overline{D}\psi$ is an F-particle. We may then consider the couple $(\psi, (1/m)D\psi)$ which is a distribution with values in the direct sum $F \oplus \overline{F}$ and define the operator $(D, -\overline{D})$ as follows:

$$(D, -\overline{D})(\psi_1, \psi_2) = (-\overline{D}\psi_2, D\psi_1).$$

We have:

$$(D, -\overline{D})((\psi, (1/m)D\psi) = m(\psi, (1/m)D\psi),$$

which is known as the Dirac equation.

If we are interested only in the representations of the proper inhomogeneous Lorentz group, it is sufficient to take $F = C^2$, and F is better than $F \oplus \overline{F}$ since it is simpler. Note that $F \oplus \overline{F}$ is irreducible under the extended spinor group, but it splits for the proper spinor group. However, if we want a partial differential equation with the intrinsic operators D and $\overline{D}$, and want to make use of representations of the covering of the extended inhomogeneous Lorentz group, it is necessary to take $F \oplus \overline{F}$ instead of F.

Vector Particles with Zero Mass

For the photon the universe is E_4 and $F = \mathbf{E}_4$. Suppose that we have determined the function $M(p)$ by choosing $M(a)$ for some point a on the light cone. But in this case, M increases infinitely at the origin and $M\mu$ is not integrable. Thus, $M\mu$ is not a measure and the Hilbert space does not exist for photons. Instead we may proceed as follows: Consider the Hilbert space for a small mass $m \neq 0$, then pass to the limit $m \to 0$, and take the resulting space as the space for mass zero.

We shall give a brief description of a particle of small mass $m \neq 0$ and spin one, and then let $m \to 0$. The universe is E_4 and $F = \mathbf{E}_4$ for particles with no charge. When $m \to 0$ we shall have a description of the photon. Let us treat the corresponding case for charged particles, that is, we complexify F to give $F = \mathbf{E}_4 + i\mathbf{E}_4$.

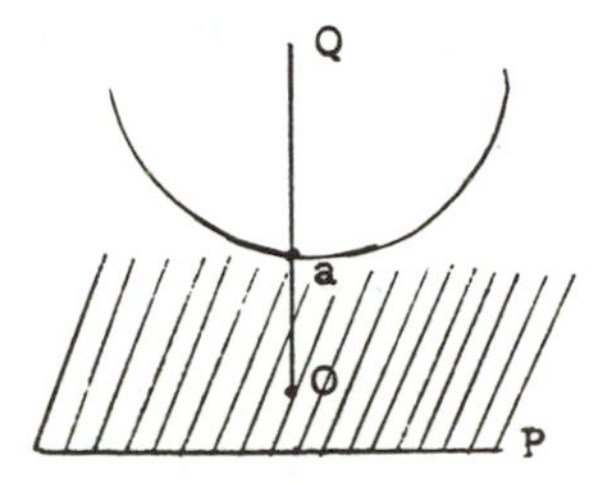

Given a point a on the hyperboloid, choose a Lorentz reference frame in which $a = (0, 0, 0, a_0)$. The stabilizer of a is the orthogonal group on the three-dimensional space-like plane P through the origin O perpendicular to the vector a. There are two independent subspaces invariant under the stabilizer:

(1) The one dimensional subspace Q spanned by the vector a.
(2) The space-like plane P.

First, consider the case where $F_a = Q + iQ$. Then for every p, F_p will be one-dimensional, and the particle will resemble a meson. When we let $m \to 0$, F_p will still be one-dimensional for every $p \neq 0$ and the particle will resemble a meson of zero mass.

Now consider the case where $F_a = P + iP$. For every p F_p is the tangent space to the hyperboloid at p and is three-dimensional. The quadratic form on F is fixed; it is the extension of the Lorentz form on $\mathbf{E}'_4$. The quadratic form on F_p is, of course, induced by the quadratic form on F and it is positive-definite because the Lorentz quadratic form is always positive definite on a space-like subspace. Let $m \to 0$. The hyperboloid is replaced by the cone. For every $p \neq 0$, F_p is the tangent plane to the cone and the quadratic form on F_p is degenerate with signature (2, 0) because it is zero on the generatrix of the cone.

The resulting space is not a Hilbert space because the norm may be zero for a non-zero element f. If $f(p)$ takes its value proportional to p at every point p, then the norm of $f(p)$ is zero. Now take the inverse Fourier transform of this space. This is a pre-Hilbert space, which is non-separated and complete. Since the space is non-separated, we have a semi-norm. In physics, a motion is defined by a class equivalence in the factor space, that is, an infinity of elements that are equivalent with respect to the semi-norm. The fact that the ψ functions equivalent with respect to the semi-norm describe the same motion is the principle of gauge invariance.

CHAPTER 6

Definition of Some Physical Notions

Scalar Case

1. *Evolution Operator.*

Let $p = (p_0, q)$ where $q = (q_1, q_2, q_3)$ is the set of the space coordinates, be the running vector of E_4', and Ω the *inferior* sheet of the hyperboloid with two sheets: $p^2 + m^2 = 0$.

Let us recall the general way of performing, in a distribution T on Ω, the change of variables:

$$q = q\,;\, u = p^2 = q^2 - p_0^2.$$

The correctness of the following writing has been justified:

$$\langle T(p^2), \varphi(p) \rangle = \int T(p^2) \quad \varphi(p) \quad dp$$

so that one may use the usual rules of the change of variables; let us compute the Jacobian:

$$\frac{\partial(u, q)}{\partial(p_0, q)} = \begin{vmatrix} -2p_0 & 2q_1 & 2q_2 & 2q_3 \\ 0 & 1 & 0 & 0 \\ 0 & 0 & 1 & 0 \\ 0 & 0 & 0 & 1 \end{vmatrix} = -\,2p_0$$

Thus: $(1/2|p_0|)\ du\, dq = dp_0\, dq$ and:

$$\int T(p^2) \quad \varphi(p) dp = \int T(u) \quad \varphi(-\sqrt{(q^2 - u)}, q) \frac{du\, dq}{2\sqrt{(q^2 - u)}},$$

since $|p_0| = \sqrt{(q^2 - u)}$, and $p_0 < 0$.

Now, let us go back to the scalar particle case; let $\mathscr{H}$ be a particle of this type. If $\psi \in \mathscr{H}$, its Fourier Transform:

$$\mathscr{F}(\psi) = \hat{\psi} = f(q) \quad \mu,$$

where μ is the density 1 spread on Ω, that is:

$$\mu = \delta(p^2 + m^2) \cdot Y(-p_0)$$

where $Y(-p_0)$ is the Heaviside Function on the lower part of E'_4. Then:

$$\langle \hat{\psi}, \varphi \rangle = \int_{E'_4} f(q) \; \delta(u + m^2) \; Y(-p_0) \, . \, \varphi(q, -\sqrt{(q^2 - u)}) \frac{du \, dq}{2\sqrt{(q^2 - u)}}$$

$$\langle \hat{\psi}, \varphi \rangle = \int_{R^3} f(q) \;\; \varphi(q, -\sqrt{(q^2 + m^2)}) \;\; \frac{dq}{2\sqrt{(q^2 + m^2)}}$$

so that, one may say that $\hat{\psi}$ is the measure on Ω defined by

$$\boxed{\hat{\psi} = f(q) \cdot \frac{dq}{2\sqrt{(q^2 + m^2)}}}$$

Proposition. Let $x = (x_0, y)$ be the set of the four variables of E_4, and $\mathscr{D}'_{x_0}(\mathscr{D}'_y)$ the set of the distributions with respect to x_0 and with values in $\mathscr{D}'_y$.

By definition $\mathscr{H} \subset \mathscr{D}'_{x_0}(\mathscr{D}'_y)$. In fact, one has:

$$\mathscr{H} \subset \mathscr{E}_{x_0}(\mathscr{S}'_y)$$

where $\mathscr{E}_{x_0}(\mathscr{S}'_y)$ is the set of the infinitely differentiable functions with respect to x_0, with values in $\mathscr{S}'_y$.

Proof. According to a general method, it is sufficient to prove it for the φ of the type $\varphi = \alpha(x_0) \cdot \beta(y)$, where $\alpha(x_0) \in \mathscr{D}_{x_0}$ and $\beta(y) \in \mathscr{D}(y)$.

$$\langle \psi, \alpha(x_0) \cdot \beta(y) \rangle = \langle \psi, \mathscr{F}\overline{\mathscr{F}}(\alpha(x_0) \cdot \beta(y)) \rangle = \langle \hat{\psi}, \overline{\mathscr{F}}(\alpha(x_0) \cdot \beta(y)) \rangle$$

Let us set $\mathscr{F}_y$ the Fourier Transform, as operating only on the three space variables. Then:

$$\langle \psi, \alpha(x_0) \cdot \beta(y) \rangle = \int_{R^3} \overline{\mathscr{F}}_y(\beta(y))\,(q) f(q) \cdot \frac{dq}{2\sqrt{(q^2 + m^2)}} \cdot$$

$$\cdot \int_{R^1} \alpha(x_0) \exp\left[-2i\pi x_0 \sqrt{(q^2 + m^2)}\right] dx_0$$

Let $\psi_{x_0}(y)$ be the distribution action *only* on the variables y, depending on the parameter x_0, and equal to $\psi(x_0, y)$.

$$\langle\psi_{x_0}(y), \beta(y)\rangle = \int_{R^3} \overline{\mathscr{F}}_y(\beta(y))\,(q)\cdot f(q)\cdot\frac{\exp\left[-2i\pi x_0\sqrt{(q^2+m^2)}\right]\,dq}{2\sqrt{(q^2+m^2)}}$$

$$\langle\psi_{x_0}(y), \beta(y)\rangle = \left\langle\frac{\exp\left[-2i\pi x_0\sqrt{(q^2+m^2)}\right]f(q)}{2\sqrt{(q^2+m^2)}}, \overline{\mathscr{F}}_y(\beta)\;(q)\right\rangle$$

$$\mathscr{F}_y\;\psi_{x_0} = f(q)\;\frac{\exp\left[-2i\pi\;x_0\;\sqrt{(q^2+m^2)}\right]}{2\sqrt{(q^2+m^2)}}$$

$$(\partial/\partial x_0)^k\;(\mathscr{F}_y\;\psi_{x_0}) = f(q)\;(-2i\pi)^k\;(\sqrt{[q^2+m^2]})^{k-1}\cdot\frac{\exp\left[-2i\pi x_0\sqrt{(q^2+m^2)}\right]}{2}$$

Hence:

$$\left|(\partial/\partial_{x_0})^k(\mathscr{F}_y\;\psi_{x_0})\right| \leqslant M_k\;\sqrt{(q^2+m^2)^{k-1}}\left|f(q)\right|$$

Thus,

$$(\mathscr{F}_y\psi_{x_0})\,(q) \in \mathscr{S}'_q, \text{ and } x_0 \to (\mathscr{F}_y\psi_{x_0})\,(q)$$

belongs to

$$\mathscr{E}_{x_0}(\mathscr{S}'_q); \text{ and } \psi = \overline{\mathscr{F}}_q\,\mathscr{F}_y\psi_{x_0} \in \mathscr{E}_{x_0}(\mathscr{S}'_y)$$

$=$ Q.E.D.

Corollary

$$\boxed{\psi_{x_0} = \psi_0 *_y \overline{\mathscr{F}}(\exp\left[-2i\pi\;x_0\sqrt{(q^2+m^2)}\right])}$$

where ψ_0 is the value of ψ_{x_0} for the particular value 0 of x_0; and $*_y$ means that the convolution operates only on the space-like variables y.

Proof. Let us consider again the preceding expression:

$$(\mathscr{F}_y\psi_{x_0})\,(q) = f(q)\cdot\frac{\exp\left[-2i\pi x_0\sqrt{(q^2+m^2)}\right]}{2\sqrt{(q^2+m^2)}} =$$

$$= (\mathscr{F}_y\psi_0)\cdot\exp\left[-2i\pi x_0\sqrt{(q^2+m^2)}\right]$$

By applying $\overline{\mathscr{F}}_q$ to both members:

$$\psi_{x_0}(y) = \psi_0(y) *_y \overline{\mathscr{F}}(\exp\left[-2i\pi x_0\sqrt{(q^2+m^2)}\right])$$

Q.E.D.

Remarks. The most general solution of the Klein–Gordon equation

$(\Box - 4\pi^2 m^2)\psi = 0$ needs two additional conditions to be determined in a unique way:

The initial values: $\begin{cases} \psi(0, y) \\ \dfrac{\partial \psi}{\partial x_0}(0, y) \end{cases}$

Instead of this, a motion ψ is characterized as soon as $\psi(0, y)$ is known. We find again the fact that the motions are very particular solutions of the Klein–Gordon equation (see end of Chap. 4).

The operator: $\overline{\mathscr{F}}(\exp[-2i\pi x_0\sqrt{(q^2 + m^2)}])*_q$

is the evolution operator.

2. *The Space $\mathscr{H}_0$ of the Heisenberg Picture.*

Let $\mathscr{H}_0$ be the set of the "initial" motions $\psi_0(y)$. Then, from the corollary, it results that:

$$\psi(x_0, y) \to \psi_0(y)$$

is a one-to-one correspondence between $\mathscr{H}$ and $\mathscr{H}_0$ (because $\psi_0(y) \equiv 0$ implies $\psi(x_0, y) \equiv 0$). On the other hand, the proposition on page 110 showed that:

$$\mathscr{H}_0 \subset \mathscr{S}'(y).$$

Proposition 1. The Hilbertian structure on $\mathscr{H}_0$ transferred from $\mathscr{H}$ by the correspondence $\psi(x_0, y) \to \psi_0(y)$ defines a topology finer than the one induced by $\mathscr{S}'_y$.

Proof. It is sufficient to write down the definitions of the various topologies involved.

Definition of H^s. To say that a distribution $T(y)$ on R^3 belongs to H^s, where s is any real number, means that:

(1) $T(y)$ is tempered;
(2) $T(q)$ is a square-summable *function* with respect to the measure: $(1 + |q|^2)^s \cdot dq$

The square of the norm on H^s is:

$$\|T\|^2_{H^s} = \int |\hat{T}(q)|^2 \cdot (1 + |q|^2)^s \cdot dq$$

Proposition 2. One has: $\mathscr{H}_0 \cong H^{1/2}$

(This implies, in particular, that the two norms are equivalent, that is: there exist two numbers A and B, fixed, such that, for every $\psi_0 \in \mathscr{H}_0$ one has $B\|\psi_0\|_{H^{\frac{1}{2}}} \leqslant \|\psi_0\| \leqslant A\|\psi_0\|_{H^{\frac{1}{2}}}$.)

Proof. The condition: $\psi_0 \in \mathscr{H}_0$ is equivalent to $\|\psi_0\|^2_{\mathscr{H}_0} < +\infty$; that is,

$$\int_{R^3} |\hat{\psi}_0(q)|^2 \; 2 \; \sqrt{(q^2 + m^2)}\, dq < +\infty$$

(cf. proof of Proposition 1) and ψ_0 is tempered. Thus $\psi_0 \in \mathscr{H}_0$ is equivalent to $\psi_0 \in H^{1/2}$.

Then, the equivalence of the two norms is obvious, since:

$$\boxed{\|\psi_0\|^2_{\mathscr{H}_0} = 2 \int_{R^3} |\hat{\psi}_0(q)|^2 \; \sqrt{(q^2 + m^2)} \; dq}$$

In brief:

$\mathscr{H}$ is the set of the solutions of the Klein–Gordon equation which verifies:

(1) *Their frequencies are negative.*
(2) *The initial condition* $\psi_0 \in H^{1/2}$.
(3) $\psi(x_0, y) \in \mathscr{E}_{x_0}(\mathscr{S}'_y)$.

3. *Position and Velocity Densities.*

The preceding expression of the square of the norm: $\|\psi_0\|^2_{\mathscr{H}_0}$ implies the following diagram is commutative:

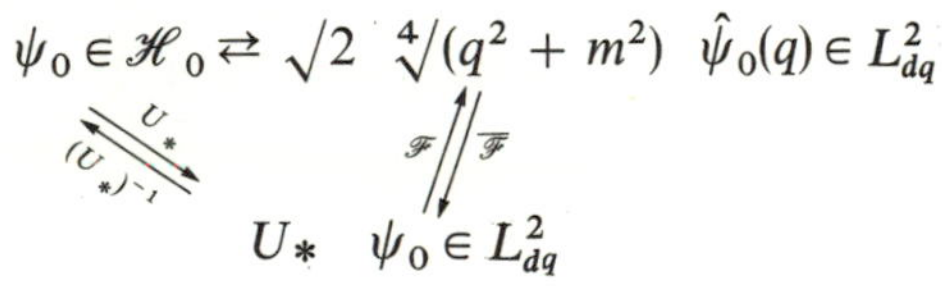

$$\begin{array}{ccc} \psi_0 \in \mathscr{H}_0 & \rightleftarrows & \sqrt{2}\,\sqrt[4]{(q^2 + m^2)}\;\hat{\psi}_0(q) \in L^2_{dq} \\ & {\scriptstyle U_*} \searrow\nwarrow {\scriptstyle (U_*)^{-1}} & \mathscr{F} \uparrow\downarrow \overline{\mathscr{F}} \\ & & U_* \;\; \psi_0 \in L^2_{dq} \end{array}$$

where $U = \mathscr{F}(\sqrt{2}\,\sqrt[4]{(q^2 + m^2)})$;

one has:

$$\|\psi_0\|_{\mathscr{H}_0} = \|\sqrt{2}\,\sqrt[4]{(q^2 + m^2)} \cdot \hat{\psi}_0(q)\|_{L^2} = \|U_*\psi_0\|_{L^2}$$

($\overline{\mathscr{F}}$ is, indeed, unitary), hence U_* is unitary.

Definition 1. I call probability of presence of the particle with the motion ψ, at the time x_0 in the volume A:

$$P_\psi(A, x_0) = \int_A |U \ast \psi_{x_0}|^2 \, dy$$

Definition 2. I call probability that the particle with the motion ψ has the speed q in the range $\hat{A} \subset \tilde{\mathbf{E}}'_3$ at the time x_0:

$$\mathscr{P}_\psi(\hat{A}, x_0) = \int_{\hat{A}} |\hat{\psi}_{x_0}(q)|^2 \; 2\sqrt{(q^2 + m^2)} \; dq$$

Hence:

$$\mathscr{P}_\psi(\hat{A}, x_0) = \int_{\hat{A}} |\exp[-2i\pi x_0 \sqrt{(q^2 + m^2)}]\, \psi_0(q)|^2 \, 2\sqrt{(q^2 + m^2)}\, dq$$

$$\mathscr{P}_\psi(\hat{A}, x_0) = \int_{\hat{A}} \frac{|f(q)|^2}{2\sqrt{(q^2 + m^2)}} dq$$

It follows:

$\mathscr{P}_\psi(\hat{A}, x_0)$ does not depend on x_0 (constant speed).
$\mathscr{P}_\psi(\hat{A}, x_0)$ depends *only* on the values of $f(q)$ in $\hat{A}$.

Vector Case

1. *Position of the Problem.*
Exactly as in the scalar case, we have still:

(1) A motion $\psi \in \mathscr{D}'_{x_0}(\mathscr{D}'_y(\mathbf{F}))$ is in fact in $\mathscr{E}_{x_0}(\mathscr{S}'_y(\mathbf{F}))$.

$$(2)\begin{cases} \mathscr{F}_y \psi_{x_0} = \mathbf{f}(q) \exp[-2i\pi x_0 \sqrt{(q^2 + m^2)}] \dfrac{1}{2\sqrt{(q^2 + m^2)}} \\[2ex] \psi_0(y) = \overline{\mathscr{F}}_q \dfrac{\mathbf{f}(q)}{2\sqrt{(q^2 + m^2)}} \end{cases}$$

where $\mathbf{f}(q)$ has the following properties:

(i) $\mathbf{f}(q) \in \mathbf{F}_p$, $p = [q, -\sqrt{(q^2 + m^2)}]$
(ii) $\mathbf{f}(q)$ is dq-measurable.
(iii) $\int_{\mathbf{E}'_3} \|\mathbf{f}(q)\|_q^2 \,(dq/2\sqrt{(q^2 + m^2)} < +\infty$

The set of these functions, with the Hilbertian structure defined by the norm:

$$\|\mathbf{f}\|^2 = \int_{\mathbf{E}_3'} \|\mathbf{f}(q)\|_q^2 \, d\mu(q)$$

where

$$d\mu(q) = \frac{dq}{2\sqrt{(q^2 + m^2)}}$$

is, let us recall it, $L^2(d\mu, R^3, \{F_p\}_p, F)$; R^3 replaces Ω because of the change of variables.

Thus, we recall that $\|\mathbf{f}\|^2$ is a continuous sum of norms, each of them being taken in a different space F_p; while, in the scalar case, $\|f\|^2$ was computed only in one space; *it* is the reason why the definitions of $\mathscr{P}_\psi(\hat{A}, x_0)$ and $P_\psi(A, x_0)$ cannot be extended immediately.

Hence, to keep, for our operators, in the vector case, the same formulas as in the scalar case, it is necessary to compute the norm of $\mathbf{f}(q)$ always in the same $\mathbf{F}_w$, independent of q, and such that $\mathbf{F}_w$ be invariant for every $\sigma \in G$. So that the vector generalization of $\mathscr{F}(U_*)$ will be an operator:[1]

$$\mathscr{F}(U)(q) = \sqrt{2}\sqrt[4]{(q^2 + m^2)} \cdot V(p)$$

where $\mathscr{L} \rightarrow Vf$ is unitary of $L^2(\mu, R^3, \{F_p\}, F)$ in $L^2(\mu, R^3, F_w)$. One may write $(Vf)(p) \quad V(p) \cdot f(p)$, where $V(p) \in \mathscr{L}(F_p; F_w)$, such that:

(1) $V(p)$ is unitary (of F_p onto F_w);
(2) The following diagram is commutative:

$$\begin{array}{ccc} F_p & \xrightarrow{V_p} & F_w \\ \downarrow \sigma & & \downarrow \sigma \\ \sigma F_p = F_{\sigma p} & \xrightarrow{V_{\sigma p}} & F_w \end{array}$$

This implies that F_w is invariant by σ, which operates unitarily in F_w.

From this diagram, we can easily define $V(p)$ if we know how to define an operator $\sigma(a, p)$ belonging to G, and operating on F, such that:

(1) Its choice depends only on the points a and p of Ω (and not, for example, on the hyperplane $x_0 = 0$).

[1] To have more explanations about this choice, see the appendix.

(2) $\sigma(a, p)\, M(a) = M(p)$.

(3) $\sigma(a, p)$ is regular (that is, analytic) everywhere, and $\sigma(a, a) = I$ (I is the identity operator).

When such an operator will be constructed, $V(p)$ will be defined by the following commutative diagram:

$$\begin{array}{ccccc} F' \text{---} M(p) \rightarrow & F_p \text{---} & \overset{\text{canonical}}{\text{injection}} & \rightarrow & F \\ \Big| & \Big| & & & \uparrow \\ \check{\sigma}(a, p) & V(p) & & & \sigma(a, p) \\ \downarrow & \downarrow & \text{canonical} & & \Big| \\ F' \text{---} M(a) \rightarrow & F_a \text{---} & \text{injection} & \rightarrow & F \end{array}$$

where $\check{\sigma}(a, p)$ is the contra-gredient of $\sigma(a, p)$, that is $V(p) = (\sigma(a, p))^{-1}$.

2. *Construction of $\sigma(a, p)$, when G/G_0 is the Lorentz group or its covering group.*

In this case, G/G_0, as set of operators in $\mathbf{E}'_4$, is known to contain only one element $\sigma_{a,p}$ such that:

(1) $\sigma_{a,p}$ transforms a fixed point a into p.

(2) $\sigma_{a,p}$ leaves invariant the two-dimensional plane through o, a, p.

(3) $\sigma_{a,p}$ operates identically in the plane Lorentz-orthogonal to the plane (o, a, p).

We take, then, for $\sigma(a, p)$, the operator in F which corresponds, according to the definition of the structure group, to $\sigma_{a,p}$. This correspondence is analytic, and thus, for proving the analyticity of $p \rightarrow \sigma(a, p)$, it is sufficient to prove the one of $p \rightarrow \sigma_{a,p}$.
Let us recall that $\mathbf{p} \in E'_4$ can be written:

$$\mathbf{p} = (p_0, \mathbf{q}) \text{ where } \mathbf{q} \in R^3, \text{ or } \mathbf{p} = p_0 \mathbf{e}_0 + \mathbf{q}.$$

Let $\mathbf{a}, \mathbf{p}$ be two elements of Ω, and $\mathbf{y}$ any element of E'_4. Since $\mathbf{a}$ is a time-like vector—indeed, it belongs to Ω—it is possible to take for $\mathbf{e}_0 = \mathbf{a}/|\mathbf{a}|$. Then, the proof consists in performing $\sigma_{\mathbf{a},\mathbf{p}}$ on $\mathbf{y}$ and checking its coordinates are transformed analytically. $\sigma_{\mathbf{a},\mathbf{p}}$ transforms the unit vector of $\mathbf{a}$ in the one of $\mathbf{p}$:

$$\sigma_{\mathbf{a},\mathbf{p}} \cdot \mathbf{e}_0 + \frac{\mathbf{p}}{m} = \frac{p_0\, \mathbf{e}_0}{m} + \frac{\mathbf{q}}{m} = \mathbf{e}_0\, ch\, \varphi + \frac{\mathbf{q}}{|\mathbf{q}|}\, sh\, \varphi$$

where

$$m = |\mathbf{p}| = \sqrt{(p_0^2 - \mathbf{q}^2)}, |\mathbf{q}| = \sqrt{\mathbf{q}^2}.$$

Hence

$$ch\varphi = \frac{p}{m}; \; sh\varphi = \frac{|\mathbf{q}|}{m}.$$

It follows that, on the bases $(\mathbf{e}_0, \mathbf{q}/|\mathbf{q}|)$, the matrix of $\sigma_{\mathbf{a},\mathbf{p}}$ is:

$$\begin{pmatrix} ch\varphi & sh\varphi \\ sh\varphi & ch\varphi \end{pmatrix} = \frac{1}{m}\begin{pmatrix} p_0 & |\mathbf{q}| \\ |\mathbf{q}| & p_0 \end{pmatrix}$$

so that:

$$\sigma_{\mathbf{a},\mathbf{p}} \cdot \frac{\mathbf{q}}{|\mathbf{q}|} = \mathbf{e}_0 \cdot \frac{|\mathbf{q}|}{m} + \frac{\mathbf{q}}{|\mathbf{q}|} \cdot \frac{p_0}{m}$$

Now let us decompose $\mathbf{y}$ in the plane $(o, \mathbf{a}, \mathbf{p})$, that is, the plane $(o, \mathbf{e}_0, \mathbf{q}/|\mathbf{q}|)$, and in its orthogonal plane:

$$\mathbf{y} = \left[- (\mathbf{y}|\mathbf{e}_0)\mathbf{e}_0 + (\mathbf{y}|\mathbf{q}) \cdot \frac{\mathbf{q}}{\mathbf{q}^2} \right] + \left[\mathbf{y} + (\mathbf{y}|\mathbf{e}_0)\mathbf{e}_0 - (\mathbf{y}|\mathbf{q}) \cdot \frac{\mathbf{q}}{\mathbf{q}^2} \right]$$

Let us compute:

$$\sigma_{\mathbf{a},\mathbf{p}} \cdot \mathbf{y}$$

By definition of $\sigma_{\mathbf{a},\mathbf{p}}$ the second bracket remains invariant, and, in using the above formulas for $\sigma_{\mathbf{a},\mathbf{p}}\mathbf{q}$ and $\sigma_{\mathbf{a},\mathbf{p}}\,\mathbf{e}_0$, we get:

$$\sigma_{\mathbf{a},\mathbf{p}} \cdot \mathbf{y} = \frac{1}{m}\left[- (\mathbf{y}|\mathbf{e}_0) \cdot p_0 \mathbf{e}_0 - (\mathbf{y}|\mathbf{e}_0)\mathbf{q} + (\mathbf{y}|\mathbf{q}) \cdot \mathbf{e}_0 + (\mathbf{y}|\mathbf{q}) \frac{p_0}{q^2} \cdot \mathbf{q} \right]$$

$$+ \left[\mathbf{y} + (\mathbf{y}|\mathbf{e}_0)\mathbf{e}_0 - (\mathbf{y}|\mathbf{q}) \cdot \frac{\mathbf{q}}{q^2} \right]$$

$$\sigma_{\mathbf{a},\mathbf{p}} \cdot \mathbf{y} = \mathbf{y} + \left[(\mathbf{y}|\mathbf{e}_0) \cdot \left(\mathbf{e}_0 - \frac{\mathbf{p}}{m} \right) + (\mathbf{y}|\mathbf{q}) \cdot \left(\frac{\mathbf{e}_0}{m} + \frac{\mathbf{q}}{q^2}\left(\frac{p_0}{m} - 1 \right) \right) \right]$$

Now, with this type of transformation, it is sufficient to check that:

$$\frac{1}{\mathbf{q}^2}\left(\frac{p_0}{m} - 1 \right)$$

is C^∞ even in $\mathbf{q} = 0$.
$p_0 = \sqrt{(q^2 + m^2)}$ so that

$$\frac{1}{q^2}\left(\frac{p_0}{m} - 1\right) = \frac{1}{m\,q^2}\left(\sqrt{(q^2 + m^2)} - m\right)$$

and for $|\mathbf{q}| \to 0$:

$$\frac{1}{q^2}\left(\frac{p_0}{m} - 1\right) \approx \frac{1}{m\,q^2}\left(m + \frac{1}{2}\cdot\frac{q^2}{m} - m\right) = \frac{1}{2m^2}$$

3. *Construction of Various Operators and Densities.*

From what we have said at the beginning of section 1, it follows that in setting:

$$\chi_0(q) = \sqrt{2}\cdot\sqrt[4]{(\mathbf{q}^2 + m^2)}\cdot\hat{\psi}_0(\mathbf{q}),$$

we have:

(1) χ_0 is dq-measurable;
(2) $\chi_0(q) \in F_q$;
(3) $\int_{R^3} \|\chi_0(q)\|_q^2 \cdot dq < +\infty$.

that is, $\chi_0 \in L^2\,(dq, R^3, \{F_p\}_p, F)$; which is exactly equivalent to:

$$f \in L^2\,(d\mu, R^3, \{F_p\}_p, F)$$

where

$$d\mu(q) = \frac{dq}{2\sqrt{(q^2 + m^2)}}.$$

Then we take:

$$\mathscr{F}(U)(q) = \sqrt{2}\ \sqrt[4]{(q^2 + m^2)}\ \ V(p)$$

where V is the operator defined by the operator $V(p)$ which satisfies to the last commutative diagram, that is $V(p) = (\sigma(a, p))^{-1}$.

Since $\sigma(a, p)$ is analytic, it is measurable, and thus $\sigma^{-1}(a, p)$ defines an operator $V(p)$ analytic in p, unitary of F_p *onto* F_a, and $\chi_0(q) \to V(p)\cdot\chi_0(q)$ is unitary from $L^2\,(dq, R^3, \{F_p\}_p, F)$ onto $L^2\,(dq, R^3, F_a)$.

Now, we get the generalization of the scalar formulas in replacing $\hat{\psi}_0(q)$ by $V(p)\ \ \hat{\psi}_0(q)$.

Fourier transform of the evolution operator is given by:

$$\hat{\psi}_{x_0} = \exp\left[-2i\pi x_0 \sqrt{(q^2 + m^2)}\right] \hat{\psi}_0(q)$$

Velocity:

$$\mathscr{P}_{\psi}(\hat{A}, x_0) = \int_{\hat{A}} \|V(p)\cdot \mathbf{f}(q)\|^2_{F_a} \cdot \frac{dq}{2\sqrt{(q^2 + m^2)}}$$

Position:

$$P_{\psi}(A, x_0) = \int_{A} \| U_* \psi_{x_0} \|^2_{F_a} \cdot dy$$

Hamiltonian (= Energy) H:

By definition it is the inverse Fourier transform of $-2\pi p_0 = 2\pi\sqrt{(q^2 + m^2)}$. Indeed, in taking the inverse Fourier transform of the both members of:

$$p_0 \cdot \hat{\psi} = \sqrt{(q^2 + m^2)} \cdot \hat{\psi}$$

we get:

$$-\overline{\mathscr{F}} p_0 * \psi = \overline{\mathscr{F}}(\sqrt{[q^2 + m^2]}) * \psi.$$

Since $\overline{\mathscr{F}}(-2\pi p_0) = i(d/dx_0)\,\delta$, this equality becomes:

$$i(\partial/\partial t)\cdot\psi = \overline{\mathscr{F}}(2\pi\sqrt{(q^2 + m^2)}) * \psi$$
$$H = \bar{\mathscr{F}}(2\pi\sqrt{q^2 + m^2})) *.$$

4. *Extension to More General Groups.*

Up to now, except in this section (2), G is supposed only to be a Lie group, admitting an invariant abelian subgroup G_0 such that G/G_0 be the Lorentz homogeneous group $\mathscr{L}$ or its covering group. Therefore, we want to extend the results of this section (2) to the following slightly more general case.

Let G be any Lie group, operating on E and $\mathbf{E}$. $\Phi^{-1}(\mathrm{I})$ the kernel of the representation of G in $\mathscr{L}(\mathbf{F}; \mathbf{F})$ and $\Gamma = G/\Phi^{-1}(\mathrm{I})$. Here, essentially only Γ occurs; and, in particular, the translations cancel.

Since $\mathscr{L}$ does not admit any invariant non-trivial subgroup, the projection $\pi_{\mathscr{L}}[\Phi^{-1}(\mathrm{I})]$ of $\Phi^{-1}(\mathrm{I})$ in $\mathscr{L}$ is either the whole $\mathscr{L}$ or the neutral element e.

First Case: $\pi_{\mathscr{L}}[\Phi^{-1}(\mathrm{I})] = \mathscr{L}$. In this case, every element $l \in \mathscr{L}$, there exists a $g \in G$ such that $\pi_{\mathscr{L}}(g) = l$, and which operates trivially on $\mathbf{F}$. Therefore $\mathbf{F} = \mathbf{F}_a$ for every p.

We may then take $\mathsf{F} = \mathsf{F}_a$; since F_a is irreducible under the stabilizer of a. F is irreducible under G. We may use exactly the same formulas as in the scalar case.

Second Case: $\pi_{\mathscr{L}}[\Phi^{-1}(\mathrm{I})] = \{e\}$. In this case, G operates faithfully in F, and $\pi_{\mathscr{L}}(\Gamma) = \mathscr{L}$. Let N be the kernel of $\pi_{\mathscr{L}}$; N is an invariant subgroup, and, according to Levy–Malcev's Theorem, valid because $\mathscr{L}$ is semi-simple there exists a subgroup $\mathscr{L}'$ of Γ, not necessarily invariant, and such that:

$\Gamma = \mathscr{L}' \times N$ (semi-direct product). According always to the same theorem, $\mathscr{L}'$ and $\mathscr{L}$ have the same Lie algebra; from which the two possibilities:

$$(\alpha)\ \mathscr{L}' = \mathscr{L}$$

It is possible to show that the necessary and sufficient condition for the existence of a position-density is:

$$\Gamma = (\mathscr{L}' \times N)/\gamma \text{ (direct product)}$$

$$(\beta)\ \mathscr{L}' \textit{ is the Spinor group of } \mathscr{L}$$

Then, it is possible to prove that the necessary and sufficient condition for the existence of a position-density is:

$$\Gamma = (\mathscr{L}' \times N)/\gamma \text{ (}\times\text{ is the direct product)}$$

where γ is the subgroup with two elements of $\mathscr{L}' \times N; \gamma = \{(e, \varepsilon); (e', \varepsilon)\}$, where ε is the neutral element of N, e and e' the two elements in $\mathscr{L}'$ which correspond to the neutral element in $\mathscr{L}$ by the canonical projection of $\mathscr{L}'$ onto $\mathscr{L}$. This condition implies that $\mathscr{L}'$ and N commute.

Hence, in both cases (α) and (β), this commutativity is necessary. This is always verified when N is finite.

The sufficiency of the conditions is trivial.

The Intrinsic Parity

Let G, the structure group of our F-particle, be the subgroup of the inhomogeneous Lorentz group which keeps the sense of the time, but not necessarily the space orientation. Let Ω be the sheet associated to the particle; for $a \in \Omega$, let G_a be the stabilizer of a; it is a maximal *compact* subgroup of G: the full orthogonal group of the subspace E_3

orthogonal to **a**. Then, $\mathbf{F}_a$, a subspace of **F**, is the space of an irreducible representation of G_a in **F**.

Let s be the symmetry with respect to the origin in $\mathbf{E}_3$. One has:

(1) $s \in G_a$; $s^2 = I$ and s belongs to the centre of G_a.
(2) G_a defines (up to a constant factor) a quadratic form, positive definite in $\mathbf{F}_a$, of which G_a is the orthogonal group.

From (1) and (2), it results (according to Schur's Lemma) that s is represented in F_a by a scalar, the square of which is 1. Thus $s = +1$ or $s = -1$. By definition, this sign is the parity of the particle. Of course, by continuity, it is independent of the chosen point a on Ω.

Instead of the group G we have dealt with up to now, let us take its covering group, which we still denote by G. Then our new G_a is the covering group of the preceding G_a. Thus, in the new G_a there are one or more different operators which operate in $\mathbf{E}_3$ as s. But they do not belong to the centre of G_a; and thus the parity no longer has meaning. It is the case for the electron. Let us recall that this concerns only the non-interacting case.

Appendix

Density of Probability of Presence of Elementary Particles*

1. Introduction: Nonrelativistic Case

In the initial, nonrelativistic theory of quantum mechanics it is assumed that the only information we have about the state of a particle, at a given time, is its wave function Ψ, a complex function on R^3 or a complex function of three coordinates x, y, z. This function is assumed to be square integrable, $\Psi \in L^2$, and moreover one assumes

$$\iiint_{R^3} |\Psi(x, y, z)|^2 \,\mathrm{d}x\,\mathrm{d}y\,\mathrm{d}z = 1. \tag{1.1}$$

Consider an observable physical quantity, taking its values in a set X. For example, the position of the particle is a quantity with values in $X = R^3$ and so is the velocity. The energy has values in $X = R$, and so on. In classical mechanics, a measurement of such a quantity is supposed to be obtainable with arbitrary accuracy, and, for a given state, the quantity has a definite value x in X. In quantum mechanics, this unlimited precision disappears. If we make a measurement of the quantity, for a particle having the wave function Ψ, we have only a probability law P_Ψ, depending on Ψ, that is, a positive measure on X, of total mass 1. Thus, if A is a subset of X, assumed to be measurable (P_Ψ), the probability that the measurement will give a result in $A \subset X$ is $P_\Psi(A)$. It is usually assumed that this probability law P_Ψ on X must be given by a spectral decomposition of the Hilbert space L^2, with respect to X. Such a spectral decomposition is defined as follows. It is a map $P: A \to P(A) = L_A^2$, where A runs over a Borel field of subsets of X, and L_A^2 is a closed subspace of L^2, with the following properties.

(a) $L_\phi^2 = \{0\}$, where ϕ = empty set of X, 0 = origin of the vector space L^2; $L_X^2 = L^2$.

(b) If A and B are disjoint subsets of X, L_A^2 and L_B^2 are orthogonal in L^2.

* Reprinted from the *Proceedings of the Fourth Berkeley Symposium on Mathematical Statistics and Probability*, 1960, University of California Press.

(c) If A is the union of a finite or denumerable family of disjoint subsets A_n, then L_A^2 is the closure of the subspace of L^2 spanned by the $L_{A_n}^2$.

Thus the probability law P_Ψ of the physical quantity under consideration must be given by

$$P_\Psi(A) = \|\Psi_A\|^2 = \iiint_{R^3} |\Psi_A(x, y, z)|^2 \, dx \, dy \, dz, \tag{1.2}$$

where Ψ_A is the orthogonal projection of $\Psi \in L^2$ on the subspace $P(A) = L_A^2$ of L^2. Axiom (a) ensures that $P_\Psi(\phi) = 0$, $P_\Psi(X) = 1$, and (b) and (c) ensure, according to Pythagoras' theorem, that P_Ψ is a completely additive set function; it is therefore, as desired, a probability law on X.

In this model the state of the particle is given by the wave function Ψ, the observable physical quantity by the spectral decomposition P, and thus a measurement of the quantity for the state of the particle is governed by the probability law P_Ψ on X, given by (1.2). There is in $L^2(R^3)$ a trivial spectral decomposition, that for which L_A^2 is the subspace of those Ψ which are zero outside A. It is regarded as the spectral decomposition associated with the observable: "position of the particle in R^3." Therefore, we have, in a measurement, the following probability for the particle to be found in the subset A of R^3

$$P_\Psi(A) = \iiint_A |\Psi(x, y, z)|^2 \, dx \, dy \, dz. \tag{1.3}$$

For this reason, $|\Psi|^2$ is the density of probability of presence. If now we look for the spectral decomposition corresponding to the first coordinate x of the particle, taking its values in R, it must be that for which, when $B \subset R$, the probability for a measurement of x to give a result in B is

$$\iiint_{x \in B} |\Psi(x, y, z)|^2 \, dx \, dy \, dz. \tag{1.4}$$

This spectral decomposition is also the spectral decomposition associated with the self-adjoint operator on L^2 "multiplication by x". Multiplication by x is also said to be the operator associated with the measurement of x.

If now we consider the evolution in time of the given particle, we shall have, at every instant t, a wave function Ψ_t, and thus a function of time having values in L^2. It will also be a function Ψ of the four variables x, y, z, t, defining for every t a function Ψ_t of the three variables x, y, z. The usual rules of quantum mechanics say that Ψ must satisfy some Schrödinger equation such as

$$i\hbar\frac{\partial\Psi}{\partial t} = H_t\Psi_t, \tag{1.5}$$

where, for every t, the H_t is a self-adjoint operator on L^2. This self-adjointness ensures, according to known properties of Hilbert spaces, that any solution of (1.5) keeps the same norm in L^2 for every t; if, for $t = 0$, it has the norm 1, which is required by (1.1), this equality remains valid for every t, and the solution defines a valid wave function for every t, and finally a valid motion of the particle. The Hamiltonian H, or the function $t \rightarrow H_t$, depends on the mechanical conditions under consideration.

2 Relativistic Case

In special relativity no distinction is made between the three space variables x, y, z and the time variable t. The universe is a space E_4, a four-dimensional *affine space*, having an associated *vector space* $\mathbf{E}_4$. We note that E_4 is not a vector space, it has no origin, and there is no sum of any two points; $\mathbf{E}_4$ is the space of vectors of E_4. If a and b are two points of E_4, then $\overrightarrow{b - a}$ is a vector, belonging to $\mathbf{E}_4$. On E_4 is given a quadratic form, with signature (3.1) measuring the "universe lengths". A physical coordinate system is an orthonormal basis of E_4, given by an origin of E_4 and four vectors of $\mathbf{E}_4$. If $x_1, x_2, x_3, x_4 = ct$ are called the corresponding coordinates of an event (an *event* is a point of E_4), the observer sees x_1, x_2, x_3 as its space coordinates and t as its time. A sense of time and an orientation of $\mathbf{E}_4$ are also given.

The complete motion of a particle will be a wave function Ψ, a complex function on E_4. For a physical coordinate system Ψ becomes a function of four variables x_1, x_2, x_3, t, and we are led back to the situation of section 1.

We shall consider that a given particle in given mechanical conditions is characterized by all its possible motions. We may assume that all these possible motions will be all the elements of norm 1 in a Hilbert space $\mathscr{H}$ of functions on E_4. For instance, in the nonrelativistic case, for a particle characterized by the Hamiltonian H, the Hilbert space $\mathscr{H}$ was formed by all the functions Ψ of four variables x, y, z, t satisfying (1.5) and belonging to $L^2_{x,y,z}$ for every t. The norm in $\mathscr{H}$ was given by

$$\|\Psi\|_{\mathscr{H}}^2 = \iiint_{R^3} |\Psi(x, y, z, t)|^2 \, dx \, dy \, dz, \tag{2.1}$$

the result being independent of t because of the self-adjointness of H_t. We can certainly not have the same kind of results in the relativistic case, because it is not Lorentz invariant.

It is an uninteresting restriction to force Ψ to be a function; we shall only assume Ψ to be a distribution on E_4, a wave distribution. Remember that a distribution Ψ is a continuous linear form on the space $\mathscr{D}(E_4)$ of the infinitely differentiable functions on E_4. with compact support. The value of Ψ on $\varphi \in \mathscr{D}$ will be denoted by $\Psi(\varphi)$ or $\langle \Psi, \varphi \rangle$. $\mathscr{H}$ will be a subspace of the space $\mathscr{D}'(E_4)$ of the distributions on E_4. $\mathscr{H}$ will also have a given structure as a Hilbert space, and we shall assume that the norm in $\mathscr{H}$ is such that convergence in $\mathscr{H}$ implies convergence in the sense of distributions. There are infinitely many choices of $\mathscr{H}$, each of which gives a possible particle in some well-defined physical situation, and all the $\psi \in \mathscr{H}$, with norm 1 in $\mathscr{H}$, give all the possible motions of such a particle in the situation considered. We are only interested in spaces $\mathscr{H} \neq \{0\}$ since we have to deal with elements of $\mathscr{H}$ of norm 1.

3. Free Scalar Elementary Particle

For a detailed proof of the formulas given here, such as (3.3) and (4.5), see Schwartz [1].

If the particle is free (no external fields), it has to be Lorentz universal, of Lorentz invariant, in the sense that a Lorentz transformation on a possible motion Ψ must give a new possible motion.

Thus we shall assume that, for any element σ of the Lorentz group and any Ψ of $\mathscr{H}$, the transformed distribution $\sigma\Psi$ also belongs to $\mathscr{H}$, and has the same norm in $\mathscr{H}$, that is,

$$\|\sigma\Psi\|_{\mathscr{H}} = \|\Psi\|_{\mathscr{H}}. \tag{3.1}$$

Note that the transformed distribution $\sigma\Psi$ is defined, for any function $\varphi \in \mathscr{D}$ which is infinitely differentiable with compact support, by

$$\sigma\Psi(\varphi) = \Psi(\sigma^{-1}\varphi) = \Psi[\varphi(\sigma x)]. \tag{3.2}$$

Therefore, σ is a unitary operator on the Hilbert space $\mathscr{H}$, and the Lorentz group G has here a unitary representation in $\mathscr{H}$. We shall define as a *free elementary particle* a free particle (thus Lorentz invariant) for which $\mathscr{H}$ is *minimal*, in the sense that no Lorentz invariant

Hilbert space $\mathscr{H}' \neq \{0\}$ contained in $\mathscr{H}$ exists except $\mathscr{H}' = \mathscr{H}$ with a proportional norm. Therefore the unitary representation of the Lorentz group G is simply an irreducible unitary representation.

What we call here the Lorentz group is the *proper inhomogeneous Lorentz group*, that is, the group of all the affine operators of E_4 onto itself, preserving the given quadratic form, on $\mathbf{E}_4$, the orientation, and the sense of time. The word *inhomogeneous* simply means that we consider affine operators of E_4 (for example, translations), and the word *proper* means that we restrict outselves to operators preserving orientation (determinant $+1$) and sense of time.

The complete list of all these Hilbert spaces $\mathscr{H} \subset \mathscr{D}'(E_4)$, Lorentz invariant and minimal, may be obtained by different techniques, all using Fourier transforms. The result is the following. Of course, for every $\mathscr{H}$, one can also take the same with a proportional norm, but we shall not distinguish them.

(a) There is one special $\mathscr{H}$, one-dimensional, all the elements of which are constant functions Ψ. It may be interpreted as the vacuum.

(b) There is a series of spaces $\mathscr{H}_1$, depending on one parameter. These cannot be physically interpreted.

(c) There is a normal series, physically interpretable. It depends on a parameter $m_0 \geqslant 0$, which may be interpreted as the rest mass of the particle, and a parameter $\pm$, which may be interpreted as the electric charge.

In this way the only particles we have found are the π-mesons, with spin 0. We find, in this way, every possible mass m_0, including 0, which is not true in nature! One can generalize and find all the known elementary particles by looking for finite-dimensional vector-valued elementary particles, for which Ψ is finite-dimensional vector-valued, that is, Ψ has a finite number of scalar components. We find here charged particles only, because we considered complex-valued wave distributions Ψ. With real-valued distributions neutral particles are obtained.

The Hilbert space $\mathscr{H}_{m_0,+}$ may be described in the following way. Consider the distribution on $\mathbf{E}_4$

$$2\pi\Delta^-_{\frac{2\pi c m_0}{h}}(\mathbf{X}) = \text{p.v.}\left[\frac{\dfrac{\pi c m_0}{2h} N_1\left(\dfrac{2\pi c m_0}{h}\sqrt{(-\mathbf{X}^2)}\right)}{\sqrt{(-\mathbf{X}^2)}}\right] Y(-X^2)$$

$$+ \frac{\frac{cm_0}{h} K_1\left(\frac{2\pi cm_0}{h}\sqrt{(\mathbf{X}^2)}\right)}{\sqrt{(\mathbf{X}^2)}} Y(X^2)\Bigg]$$

$$+ i\left[\frac{\frac{\pi cm_0}{h}\varepsilon(X_0)J_1\left(\frac{2\pi cm_0}{h}\sqrt{(-\mathbf{X}^2)}\right)}{\sqrt{(-\mathbf{X}^2)}} Y(-\mathbf{X}^2) - \frac{1}{2}\varepsilon(X_0)\,\delta(\mathbf{X}^2)\right] \tag{3.3}$$

In this rather complicated formula N_1 is a Neumann function; K_1 a Kelvin function; J_1 a Bessel function (one could use a shorter formula with Hankel functions); Δ^- is the name of the distribution, one of the "singular functions", that is, distribution of quantum mechanics; p.v. means Cauchy's principal value; $\mathbf{X}^2$ means the value on the vector $\mathbf{X} \in \mathbf{E}_4$ of the Lorentz quadratic form; Y means the Heaviside function where $Y(\tau) = 1$ for $\tau \geqslant 0$, and $= 0$ for $\tau < 0$; ε is defined as the function $\varepsilon(\tau) =$ sign of $\tau = +1$ for $\tau \geqslant 0$, and -1 for $\tau < 0$, so that if X_0 is the fourth component of $\mathbf{X}$ in any coordinate system X_1, X_2, X_3, X_0 then $\varepsilon(X_0)$, for elements $\mathbf{X}$ of the interior or the surface of the light cone, is $+1$ for $\mathbf{X}$ in the positive light cone, -1 for X in the negative light cone; $\delta(\mathbf{X}^2)$ is defined from the $\delta(u)$ of one variable u by the change of variables $u = \mathbf{X}^2$ (we denote here distributions in the physical way, as functions); c is the velocity of light; h is Planck's constant. The parameter is written cm_0/h so that m_0 may be interpreted as rest mass of the particle. Then a distribution Ψ on E_4 belongs to $\mathscr{H}_{m_0,+}$ if and only if the expression

$$\frac{|\langle \Psi, \varphi\rangle|}{\langle 2\pi\Delta^-_{\frac{2\pi cm_0}{h}} * \overline{\varphi}, \varphi\rangle^{1/2}} \tag{3.4}$$

is bounded when φ runs over $\mathscr{D}(E_4)$. Here * means convolution. In this case the upper bound is the norm of Ψ in $\mathscr{H}_{m_0,+}$.

All the Ψ of $\mathscr{H}_{m_0,+}$ are solutions of the Klein–Gordon equation

$$\Box\,\Psi - \frac{4\pi^2c^2m_0^2}{h^2}\Psi = 0. \tag{3.5}$$

This equation is here *not assumed*; we find it as a *consequence* of our hypothesis that $\mathscr{H}$ is Lorentz invariant and minimal.

The Hilbert space $\mathscr{H}_{m_0,-}$ is obtained in the same way from Δ^+, which is obtained from Δ^- by changing i into $-i$.

4. Density of Probability of Presence

From now on we shall write $\mathscr{H}$ instead of $\mathscr{H}_{m_0, \pm}$. Then every Ψ of $\mathscr{H}$ is a priori a distribution. Actually one can prove it is a function, that is, a locally integrable function defined almost everywhere on E_4.

Consider a physical coordinate system. Thus Ψ becomes a function of (x, y, z, t), locally integrable, defined almost everywhere. Therefore, if we fix the time $t = t_0$, then Ψ is not *defined* as a function of x, y, z, since a hyperplane $t = t_0$ is a set of measure zero in E_4. But one can prove the following result: it is possible to choose Ψ (initially defined only almost everywhere) so that it is a *continuous function* of t, with values in the space L^1_{loc} of the locally integrable functions of (x, y, z). Because of the continuity in t, the function Ψ is then determined not merely almost everywhere in E_4 but, *for every* t, almost everywhere with respect to (x, y, z).

Finally, Ψ defines for $t = t_0$, a well-defined Lebesgue class of functions Ψ_{t_0}, and also a well-defined distribution Ψ_{t_0} on R^3. Moreover, it can be proved that a knowledge of Ψ_{t_0}, *the cross section of* Ψ *over the hyperplane* $t = t_0$, completely determines Ψ (quantum-mechanical determinism). The system of the function Ψ_{t_0} is a subspace $\mathscr{H}_{t_0}$ of $\mathscr{D}'(R^3)$, having a one-to-one correspondence $\Psi \to \Psi_{t_0}$ with $\mathscr{H}$. Carrying over the Hilbert structure of $\mathscr{H}$ onto $\mathscr{H}_{t_0}$, we define $\mathscr{H}_{t_0}$ as a Hilbert space contained in $\mathscr{D}'(R^3)$, which may be called the cross section of the Hilbert space $\mathscr{H}$ by $t = t_0$. Now any physical observable quantity at the time t_0, with values in a set X, must be measured by a spectral decomposition of $\mathscr{H}_{t_0}$, relative to X. If $A \to P(A) = (\mathscr{H}_{t_0})_A$ is this spectral decomposition, the probability of finding the value of a measurement of the quantity in A, when the wave function is $\Psi \in \mathscr{H}$, with $\|\Psi\| = 1$, will be

$$P_\Psi(A) = \|(\Psi_{t_0})_A\|^2, \tag{4.1}$$

where $(\Psi_{t_0})_A$ is the orthogonal projection of Ψ_{t_0} on $(\mathscr{H}_{t_0})_A$. Since $\|\Psi_{t_0}\| = \|\Psi\|$, where the Hilbert structure on $\mathscr{H}_{t_0}$ is defined by carrying over that of $\mathscr{H}$, we have that P_Ψ is, as desired, a probability law on X. We are interested in the measurement of the position of the particle at the time t_0, whose physical quantity, the position, has values in R^3. Here the result is essentially different from that of the nonrelativistic case. One cannot postulate that the manifold $(\mathscr{H}_{t_0})_A$ is formed by all the Ψ equal to zero outside A, because, as is seen by studying the scalar product in $\mathscr{H}_{t_0}$, in this case $(\mathscr{H}_{t_0})_A$ and $(\mathscr{H}_{t_0})_B$ would not be orthogonal subspaces in $\mathscr{H}_{t_0}$. In other words, $|\Psi_{t_0}|^2$ cannot be the density of probability of presence. In yet other words, the "position operator" in coordinate x_i for $i = 1, 2, 3$, cannot be multiplication by x_i, as it is in the nonrelativistic case, because such an operator is not self-adjoint in the Hilbert space $\mathscr{H}_{t_0}$. In the physical literature a density of probability of presence for the meson is often considered which is not even positive! What should be the spectral decomposition relative to R^3, corresponding to the measurement of position at the time t_0?

It is natural to ask whether there exists a one-to-one norm preserving, linear transformation $\Psi_{t_0} \to \oplus$, from $\mathscr{H}_{t_0}$ onto $L^2(R^3)$, that is, covariant with the inhomogeneous proper orthogonal group Γ of R^3. That is, Ψ_{t_0} must be such that

$$\Psi_{t_0} \to \oplus \text{ implies } \tau\Psi_{t_0} \to \tau\oplus \tag{4.2}$$

whenever τ belongs to Γ. Note that Γ is the group of affine operators of R^3, preserving lengths and the orientation. Here inhomogeneous means that it contains the translations, proper that it preserves the orientation.

In this case the trivial spectral decomposition of $L^2(R^3)$ will define a spectral decomposition of $\mathscr{H}_{t_0}$ and $(\mathscr{H}_{t_0})_A$ will be the set of $\mathscr{H}_{t_0}$ corresponding to the set of L^2 formed by all the $\oplus$ equal to zero outside A. Such a spectral decomposition will be acceptable as a spectral decomposition for the measurement of the position of the particle at the time t_0, and $|\oplus|^2$ will be acceptable as a possible density of probability of presence at the time t_0, for the particle having the wave function Ψ or the instantaneous t_0-wave function Ψ_{t_0}.

In fact, such a map $\Psi_{t_0} \to \oplus$ can be found. It is given as follows. If Δ is a Laplacian on R^3, by Fourier transform $\mathscr{F}$ there is classically defined an operator

$$\sqrt{2}\left(-\frac{\Delta}{4\pi^2} + \frac{c^2 m_0^2}{h^2}\right)^{1/4}. \tag{4.3}$$

Thus, one has

$$\oplus = \sqrt{2}\left(-\frac{\Delta}{4\pi^2} + \frac{c^2 m_0^2}{h^2}\right)^{1/4} \Psi \tag{4.4}$$

or

$$\mathscr{F}\oplus = \sqrt{2}\left(\rho^2 + \frac{c^2 m_0^2}{h^2}\right)^{1/4} \mathscr{F}\Psi_{t_0}, \tag{4.5}$$

where ρ is the distance from the origin.

Actually, it may be written as a convolution,

$$\oplus = \Psi_{t_0} * \frac{\left[2^6\left(\frac{cm_0}{h}\right)^7 \pi^{-1}\right]^{1/4}}{\Gamma\left(-\frac{1}{4}\right)} \rho^{-7/4} K_{7/4}\left(2\pi \frac{cm_0}{h}\rho\right), \tag{4.6}$$

where K is a Kelvin function, decreasing, classically, exponentially at infinity. As we observe, $\oplus$ is obtained from Ψ_{t_0} by a convolution, which is a *nonlocal* operation. Therefore, knowledge of Ψ_{t_0} in an open set Ω of R^3 does not allow us to know $\oplus$ in Ω; for this, a complete knowledge of Ψ_{t_0} is necessary.

There are infinitely many other isometries of $\mathscr{H}_{t_0}$ onto $L^2(R^3)$ having the same property of covariance with the orthogonal group. Namely, one can take the previous one followed by any unitary transformation of L^2 onto itself, commuting with the inhomogeneous proper orthogonal group of R^3. Such a unitary transformation $\oplus' \to \oplus''$ is given, using the Fourier transform $\mathscr{F}$, by the formula

$$\mathscr{F}\oplus'' = (\mathscr{F}\oplus')e^{if(\rho)}, \tag{4.7}$$

where $f(\rho)$ is an arbitrary measurable function of the distance ρ from the origin of R^3.

Therefore, all the possible operations $\Psi_{t_0} \to \oplus$ are of the form

$$\oplus = \mathscr{L} * \Psi_{t_0}, \tag{4.8}$$

where

$$\mathscr{F}\mathscr{L} = \sqrt{2}\left(\rho^2 + \frac{c^2 m_0^2}{h^2}\right)^{1/4} e^{if(\rho)}. \tag{4.9}$$

Since the coefficient of $e^{if(\rho)}$ is real and nonnegative, while $e^{if(\rho)}$ itself is never real and nonnegative unless $f(\rho) = 2k\pi$, it can be seen that there is one and only one transformation of the form (4.9), where $\mathscr{L}$ is a distribution of positive type, having a positive measure as Fourier transform. But I do not see any physical reason for $\mathscr{L}$ to be of positive type.

I should rather think that in the correspondence between the physical particle and the mathematical representation, there remains some arbitrariness. One example is the choice of $\mathscr{L}$, and the simplest choice is given in (4.6). The same can be done for vector-valued (spin) particles.

REMARK. Of course, the formulas and equations given here are well known in physics; only the point of view and the method of exposition are new (and, eventually, the mathematical rigor!).

Our density of probability of presence was already introduced by Newton and Wigner [2].

References

[1] L. SCHWARTZ, *Matemática y Física Cuántica*, Buenos Aires, Universidad de Buenos Aires, 1958.

[2] T. D. NEWTON and E. P. WIGNER, "Invariant theoretical determination of position operators", *Phys. Rev.*, Vol. 76 (1949), p. 191(A).

Index

QUEEN MARY
COLLEGE
LIBRARY

WESTFIELD CO UNIV. LONDON